Materials: A Very Short Introduction

VERY SHORT INTRODUCTIONS are for anyone wanting a stimulating and accessible way in to a new subject. They are written by experts, and have been translated into more than 40 different languages.

The Series began in 1995, and now covers a wide variety of topics in every discipline. The VSI library now contains over 350 volumes—a Very Short Introduction to everything from Psychology and Philosophy of Science to American History and Relativity—and continues to grow in every subject area.

Very Short Introductions available now:

Available soon:

For more information visit our website

www.oup.com/vsi/

Christopher Hall

MATERIALS

A Very Short Introduction

OXFORD
UNIVERSITY PRESS

OXFORD
UNIVERSITY PRESS

Great Clarendon Street, Oxford, OX2 6DP,
United Kingdom

Oxford University Press is a department of the University of Oxford.
It furthers the University's objective of excellence in research, scholarship,
and education by publishing worldwide. Oxford is a registered trade mark of
Oxford University Press in the UK and in certain other countries

Published in the United States of America by Oxford University Press
198 Madison Avenue, New York, NY 10016, United States of America

British Library Cataloguing in Publication Data
Data available

Library of Congress Control Number: 2014942176

ISBN 978-0-19-967267-7

Printed and bound by
CPI Group (UK) Ltd, Croydon, CR0 4YY

Contents

Preface

Useful matter is a good definition of materials. It covers engineering stuff like steel, concrete, rubber, plastics, wood, semiconductors, glass, and aluminium. But it stretches to oil and gas, food, agrochemicals, pharmaceuticals, explosives, textiles, and dyestuffs. It also has room for oddball things like ivory and invar, graphite, grease, porcelain, and paint. So it should. All these materials are useful to us. Materials are materials because inventive people find ingenious things to do with them. Or just because people use them.

The link between materials and users takes us beyond science and engineering, into medicine, and on into economics, history, and culture as well. Materials flow through the world economy in prodigious amounts, and commodity markets fix the prices of oil and gold. Industries grow and die; steel production moves from Europe and North America to Asia. A development economist describes the scavenging of waste plastic in India's informal economy. Anthropologists analyse material culture, and how pots and solar lamps work in societies and groups. The archaeological record consists mostly of material objects. Materials shape communications, the media, architecture, building, and the fine arts.

Most new technology depends in some way on innovation in materials. All advanced economies spend vast sums of money on materials research. Materials production is the business of big industries like steel and semiconductors. But it was so in all historical periods: from the Silk Road to Silicon Valley. After the Second World War the importance of materials in industry and economics was as clear as day. The universities were prodded by industry labs at General Electric and Bell Telephone, and by the military. Materials science appeared as an academic discipline in the 1960s, first in the USA at Northwestern University, then at Penn State, and rapidly after that in universities worldwide. Materials science today explains how materials are made and how they behave as we use them.

But this book is not a synopsis of materials science, although you will see the kind of explanations of material behaviour that materials science provides. I am taking a broader view. Even scientifically, there is more to materials than materials science. Engineering has been and is the source of much of what we know about materials, for example how we use them in large-scale structures like bridges. In the 19th century engineers and mathematicians together worked out how to describe the strength of materials. Industrial manufacturing in iron and steel, in pottery, in glass and in textiles provided more new knowledge. Much has come from chemists working on every element of the Periodic Table, and since the 1920s from physicists who brought quantum mechanics to bear on the properties of solids. And from others who invented scientific instruments.

I start with a mixer in Chapter 1 to introduce materials of many kinds, from scientific, industrial, and sometimes historical viewpoints. Chapter 2 parachutes down into the microscopic architecture of materials to see how they are constructed (and how we know). Chapters 3 and 4 describe the diverse properties of materials that make them useful, and to some extent how we explain these properties. In Chapter 5, I look at how we make materials, and how we make things from them, which is ultimately what we care about. And finally in Chapter 6, I try to make sense of the problem of sustainability.

I am grateful to many people for helping with comments and advice. In particular I thank David Coates, Ken Entwistle, Simon Finch, Mike Goulette, Ben Hall, Liza Hall, Andrew Harrison, Bill Hoff, Ying Jiang, Vassilis Koutsos, Geoff Maitland, Pablo Maldonado, Richard Nelmes, Elio Raviola, and Nick Rowley.

List of illustrations

Materials

Chapter 1
Gold, sand, and string

All about us is an uncountable profusion of materials. Gold, sand, and string are just three random picks, but they are wonderfully dissimilar. And these three materials can stand also for the metallic, the inorganic, and the organic resources on which we draw. Gold leads us to bronze, and to iron; sand to stone, to clay, and eventually to silicon; string to rope, and cloth, wood, and rubber.

It is from the particularities of substances that uses arise. So, first to gold.

Old money

Pure gold coins, stamped with a lion and a bull, were made in Anatolia in the 6th century BCE under the rule of King Croesus (Figure 1).

This marks an important moment in the invention of coinage. These croeseids, looking like jelly beans by Fabergé, each weigh about 8 grams, and each is composed of about 25 billion trillion atoms of gold. Because the element gold has only a single stable isotope, ^{197}Au, every one of these atoms is identical in structure and mass. In making the coins, the gold is melted, and on cooling the atoms which are in complete disorder in the molten metal

1. Gold croeseid, 20 millimetres long

settle into a regular solid arrangement which we describe as crystalline. An X-ray diffraction experiment shows the unmistakable signature of crystallinity, and the diffraction pattern tells us precisely how the atoms are arranged. Gold, like copper and many other metals, has a cubic structure—more precisely, a face-centred cubic (fcc) structure—with atoms close packed in sheets, the sheets stacked one above the other (Figure 2). Knowing the wavelength of the X-rays, we can calculate accurately the size of the unit cube (the building block of the regular arrangement), the edge of which for gold is 0.41 nanometre (1 nanometre is one-millionth of a millimetre: this and other units and quantities are listed at the end of the book).

The solid gold does not appear as a clip-art faceted crystal because as the molten metal cools crystallization starts simultaneously at many places. The solidified gold is an aggregate of myriads of microscopic crystallites, all firmly joined together but each differently and randomly oriented. This microstructure of crystalline grains can be revealed in gold as in other metals by etching the surface chemically. Such metallographic techniques were not known until the 19th century.

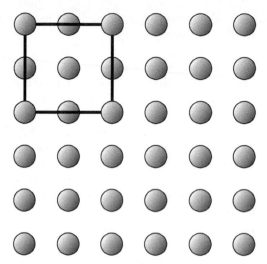

2. Gold atoms in the face-centred cubic structure. A section through the lattice, showing the unit cube, with atoms at each corner and at the centre of each face: a view along one of the cube edge directions

The Anatolian gold came from the silt of the river Pactolus, near the city of Sardis, where it gathered as it washed out from the hills above. Unlike the coins, the alluvial gold was not pure but contained about 25 per cent by weight of silver. It was evidently not only coinage that was invented at that time, but also the process of gold refining, and the two facts are connected. A coinage metal must be of warranted composition to have reliable value, and by issuing coins of pure gold and pure silver this is what Croesus achieved. We know from archaeological evidence how the parting of the gold and silver was done: native gold was hammered into thin foils and placed between layers of salt mixed with crushed pot in a sealed earthenware jar. The jar was heated on the fire to a temperature just below the melting point of gold, which is 1,064°C (degrees Celsius). The salt fumes (strictly, sodium chloride vapour) attack the silver, which is more reactive than the gold, producing silver chloride which vaporizes and is absorbed by the crushed pot. This cementation process for gold

3

refining was widely employed for many centuries, but no earlier use is known than at Sardis.

The native gold–silver material is an early instance of a solid-solution alloy. Pure silver also has a face-centred cubic structure, and it happens that the interatomic spacing is identical to that of gold. A molten mixture of gold and silver therefore effortlessly adopts the fcc structure on solidifying, with gold and silver atoms occupying positions at random in the three-dimensional solid lattice. A puzzle is that during parting the pure gold which is formed becomes riddled with tunnels and rough to the touch. These tunnels are enormously larger than the atom-sized cavities left behind by individual silver atoms. It seems that as silver is lost, the microstructure of the metal is dynamic, and continuously reorganizes itself. In fact, small cavities move and combine to form the larger tunnels. The end-product is coarse and porous, but it is also malleable, and by burnishing and planishing a goldsmith brings it to a bright, smooth finish.

Besides silver other elements are often found in ancient gold. Copper and iron can also dissolve in gold, and may be present in small concentrations. Egyptian rose-gold contains 10 per cent or more of copper, and this imparts a distinctive colour. Small amounts of platinum-group metals, usually osmium and iridium, may also occur. These do not dissolve, but exist as separate particles embedded in the gold matrix. Sensitive analytical methods also detect many other elements at trace levels. Such low concentrations have no effect on the appearance or properties of the gold but may act as a fingerprint to give clues to its source. So, the abrupt fall in the tin content of Spanish gold coins in the middle of the 8th century reveals the arrival of West African gold in Spain after the Arab conquest.

The history of Anatolian gold coins shows a material acting as the pivot of economic and social changes. I have described it from the viewpoint of a materials scientist, using this example to give a technical framework which can be adapted to other metals, and even to non-metallic materials as well. We understand all materials by knowing about composition and microstructure. Despite their extraordinary minuteness, the atoms are the fundamental units, and they are real, with precise attributes, not least size. Solid materials tend towards crystallinity (for the good thermodynamic reason that it is the arrangement of lowest energy), and they usually achieve it, though often in granular, polycrystalline forms. Processing conditions greatly influence microstructures which may be mobile and dynamic, particularly at high temperatures. Structural features may develop at coarser scales than the atom, or the crystal lattice or the grain, as in the tunnel porosity of de-silvered gold.

Bronze bells

The *metallic tradition* (the phrase is J. E. Gordon's) has dominated engineering over recent centuries, and led to the combustion engine, the oil tanker, the turbine, the aeroplane, the suspension bridge, the skyscraper, the petrochemical plant, and the machines of war. It developed of course not from gold and silver and jewellery and coins, but from metals like copper and iron with greater technological clout. Copper ores (oxides and sulfides mostly) if not abundant are widely distributed and conspicuously colourful. Even native metallic copper is sometimes found. In the ores, the copper is only weakly combined with oxygen and sulfur, so the metal can be freed through simple smelting by heating with charcoal. Copper metal melts just below 1,100°C, only a few degrees higher than gold, within reach of primitive pyrotechnology and the leather bellows. So molten copper was poured and cast. But it was in bronzes, alloys formed by mixing copper first with arsenic and later with tin, that copper had a profound technological impact from the 3rd millennium BCE and in many places. While pure copper is a soft and malleable metal, the bronzes are strong and fit for new uses, particularly for making tools and weapons.

And bells. Bronze bells were made in China in the Zhou dynasty of the 2nd millennium BCE, and were of ritual significance in the time of Confucius (Figure 3).

Much later, around the 12th century CE, a bell-making tradition emerged in Europe. It is remarkable that both Chinese and European bell metals were tin bronzes of similar alloy composition (about four parts copper to one part tin by weight); and equally that there is little difference in composition between modern bell metals and those of the 65 chime bells of Marquis Yi of Zeng who died in 433 BCE. The bell developed as a creative expression of the material properties of tin bronze (as Michelangelo claimed to liberate his sculpture from the marble). Interlaced are the making-by-casting, the toughness of the cast bell which does not

3. **Chinese chime bell**

crack when struck, and above all the musical and acoustic qualities of the material. Unlike the circular European bells, the Chinese chime-bells were oval, and produced different tones when struck at two different points. Much later, in 1890, the acoustics of the bell was explained scientifically by Lord Rayleigh, who showed that when a bell is rung many modes of vibration are excited simultaneously, and that these decay away at different rates. The lowest tone (the hum) of a large bell may be audible for up to a minute after the strike. The bell sound depends on a favourable set of material attributes found only in bronze bell metal. The speed of sound through the solid bronze is one of these, and another, more subtle, is the slow decay of the various vibrations, a property which depends on peculiarities of the polycrystalline grain structure. Unlike gold and silver, copper and tin are not isostructural. The bell-metal alloy is not a simple solid solution because copper can dissolve only at most about 10 per cent of tin, so the balance of the tin combines with copper to form a second kind of crystallite, approximately Cu_3Sn_8 in composition. This kind of fixed-composition intermetallic phase is common in alloys. Underlying the acoustical properties is the multiphasic structure of bronze bell metal (sometimes called Chinese high-tin bronze), which is sensitive to both the composition and the temperatures experienced during casting and cooling.

The finely tuned bell-metal bronzes were the forerunners of a clutch of other copper alloys. Zinc-copper bronze, known by long usage as brass, was much used in ships because of its indifference to sea water. Naval brass (and admiralty brass) was and is the stuff of marine propellers, and even sailors' buttons. Copper is an easy metal for alloying: now we have also nickel bronze, aluminium bronze, silicon bronze, beryllium bronze, and phosphor bronze, all workhorse materials for mechanical engineering.

The tenfold increase in copper production during the second half of the 19th century had little to do with buttons. It happened because electrical engineers found uses for copper wire. In 1831 when Michael Faraday made his induction ring (the first electrical transformer) he wrapped around the iron core 130 feet of pure copper wire from his laboratory. It was already known that soft pure copper could be drawn into a flexible wire by pulling through a small hole, but that was of no great value until electrical machines were invented. Faraday's wire was made for mechanical bell-pulls, and he insulated it himself with calico and string. Today 60 per cent of world copper production (say, 10 million tons a year) goes into motor windings, telephone and electricity cables, vehicle electrical systems, and consumer electronics. An early, transformative project was the submarine cable across the Atlantic. For this in 1857 William Thomson worked out the physics of sending an electrical signal down a long wire ('It is the most beautiful subject possible for mathematical analysis'). With his students in Glasgow he measured the conductivity of copper. It was high but variable. Higher conductivity went with higher purity. Trace arsenic or antimony was often the culprit. Commercial copper wire now is 99.99 per cent pure (1 impurity atom in 10,000), made by dissolving unrefined copper in sulfuric acid and redepositing it on the cathode of an electrical cell. The early submarine cables were difficult to lay, did not work well, and sometimes failed entirely. The benefits of being able to rely on materials with designated properties were clear to the Board of Trade enquiry. Henry Fleeming Jenkin and others had to invent better methods for measuring conductivity (or resistivity), and give precise meaning (and names) to electrical units such as the ohm. Thus the submarine telegraph and the problem of copper conductivity drove developments in metrology, with wide impact both in industry and in fundamental science.

Big steel

Pure iron melts at 1,536°C, more than 400°C higher than copper. Such temperatures were out of reach before the industrial steelmaking of the 19th century. Many centuries of artisan iron working were thereafter displaced by the giant plants of steel cities like Pittsburgh, and later by others such as Nowa Huta and Anshan. World steel production now exceeds one billion tons per year, 20 times that of aluminium and copper combined. Of course, in earlier centuries, iron was laboriously worked without melting. The fact that steel is an alloy of iron and carbon was not known. Even so, the hard-won skills of smelting, forging, quenching, tempering, and carburizing were successful in producing wrought-iron artefacts on a small scale, some of great quality and beauty. A little cast iron was made directly from the blast furnace. But it was large-scale steelmaking using the Bessemer converter that first made cheap steel for railways, bridges, and large buildings. Big steel made the skyscraper possible, although it did not guarantee that it would happen, and in most countries it didn't. Later, Bessemer was replaced by the open-hearth and then the basic oxygen process, and by the electric arc furnace. All make molten steel at white heat by the ton.

Once molten-metal steelmaking arrived (say from the 1850s), surprising technical discoveries came in spate. Some were to do with plain carbon steels, and how to improve and control properties by heat treatment. How to roll steel, how many times, at what temperatures, and how quickly to cool it. Many hard lessons were learned when production of Bessemer steel was ramped up to feed the building of the railways. A lot of rail was made, but rails broke too often and in the end Bessemer steel was so poorly regarded that engineers demanded open-hearth steel for buildings and bridges. But gradually it became clear how to control and predict the properties of steels made for different purposes. Just as for copper, it was obvious that standards and specifications were needed, and for steel it was the car makers who worked them out.

Other discoveries were about alloys of iron with other metals. In 1882 Robert Hadfield in Sheffield found that adding manganese produced a strong, tough steel that was good for railway wheels (and then for tanks); and 30 years later Harry Brearley, another Sheffielder, discovered that adding chromium made a steel that did not rust. In both, the alloy additions were more than 10 per cent. These and other alloy steels, often made in small batches, are exploited in umpteen uses. At Bethlehem Steel Co around 1900 Frederick W. Taylor's 8 per cent tungsten tool-steels kept their cutting edge for many hours even when running at red heat. Tungsten steels were not new, but a vital part of Taylor's invention was a high-temperature heat treatment so elaborate and exact that it took him years to perfect. Taylor was driven by a passion for efficiency. His 'high-speed' tool steels transformed manufacturing practice in every industry.

The science of ferrous alloys developed rather separately and slowly. There was no big breakthrough. New methods of chemical analysis were important. The picture eventually emerged that carbon was the key to understanding iron, but the science did not create big steel. The bare bones are: that brittle pig iron from the blast furnace has a high carbon content, perhaps 4–5 per cent by weight; that removing part but not all the carbon converts pig iron to steel; that steel properties depend both on carbon content and on heat treatment. This picture of the iron–carbon system needed three legs under it. First, a workable atomic theory, with notions of atomic weight and size; then, some evidence of what iron alloys look like microstructurally; and finally an idea of what chemical reactions take place in iron- and steelmaking.

The pieces came together to show that the high temperature crystalline phase which forms when iron first solidifies has an fcc structure (just like gold, shown in Figure 2). It is known as gamma-iron. Gamma-iron can dissolve a lot of carbon, up to about 2 per cent by weight. Since the iron atom is 4.5 times heavier than the carbon atom, this means that almost one in ten of the atoms in the gamma-iron solid solution may be carbon. The radius of the C atom is little more than half that of the Fe atom, so the solid solution is not like that in gold–silver, where a silver atom replaces a gold atom on a lattice site. In Fe–C all the Fe atoms remain in place but the C atoms sit between them, under the skirts of the larger Fe atoms, to form an interstitial solid solution. The fcc structure has convenient cavities for the C atoms to occupy. So far so good. The drama of steel arises because as we continue to cool (in the rolling mill or on the blacksmith's anvil) the gamma-iron transforms to a different crystal structure (alpha-iron) *which cannot dissolve carbon*. Alpha-iron has a body-centred cubic (bcc) structure (Figure 4), subtly different from the fcc structure of gamma-iron. If the cooling is slow enough, the carbon migrates out of the bcc crystals to form a new phase, which is a fixed-composition intermetallic (like that formed in bell-metal bronze), in this case with the composition Fe_3C, called cementite. The solubility of carbon in the bcc phase is almost zero, so a steel of this kind is a bcc alpha-iron, almost pure, in which cementite is dispersed and embedded. With a small overall C content, say 0.2 to 0.5 per cent, this is a plain carbon or

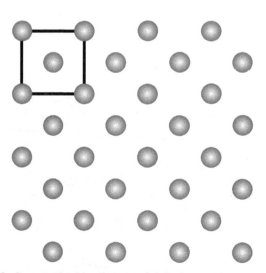

4. **The body-centred cubic structure of alpha-iron, with atoms at each corner and at the centre of the unit cube**

mild steel. On the other hand, if we take the hot gamma-iron and quench it into a bath of oil, the C atoms do not have time to migrate out of the fcc phase and are trapped. Because of the trapped carbon, the structural gamma–alpha phase change is obstructed. A possible outcome is that the cubic fcc structure glides into a distorted *martensite* version of the bcc structure which can accommodate the carbon. We have made martensitic steel with seriously different properties from the slow-cooled (pearlitic) steel. The martensitic transformation by non-diffusive gliding of atoms (and subject to some severe crystallographic rules) was first observed in iron and steel, but occurs widely in materials of many kinds, even in biology. Between the very-slow and very-fast versions, we have most of steelmaking: cooling, annealing, quenching, all tweaked by minor and major alloy additions, and tweaked again by mechanical operations like rolling.

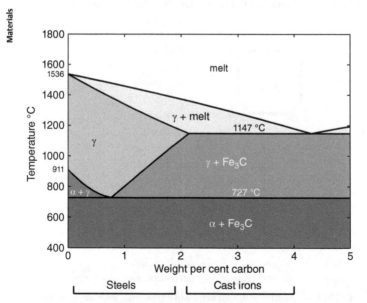

5. **A simple form of the iron–carbon equilibrium diagram**

The relations between composition and temperature are displayed in the map (Figure 5) first drawn by William Roberts-Austen in 1897 (a few months before the death of Henry Bessemer), and more correctly by H. W. Bakhuis Roozeboom three years later. It is the first and best-known example of a metallurgical phase diagram, thousands of which have since been constructed and which guide (often retrospectively) the interpretation of alloy behaviour.

Quartz, calcite, and clay

Oxygen is the most abundant element in the earth's crust and silicon the second. In nature, silicon occurs always in chemical combination with oxygen, the two forming the strong Si–O chemical bond. The simplest combination, involving no other elements, is silica; and most grains of sand are crystals of silica in the form known as quartz. It is easy to see that quartz is crystalline because we often find it as a faceted crystal. The paradox is that quartz crystallizes only with difficulty. It melts (at well above 1,500°C) to form a treacly liquid which on cooling behaves quite unlike the metals. While metals crystallize easily as they cool, the entangled liquid of silicon and oxygen atoms, still mostly bonded together, do not find the long-range ordered positions that define the crystal structure. The disorder of the liquid is largely preserved in a material which in all practical respects is a solid. We have formed a glass. So why are quartz crystals familiar to us? They come from two sources. Natural quartz is formed either in molten rocks which cool extremely slowly or else in cooler, wet conditions, also extremely slowly. In water, silica dissolves (only slightly but enough) and then quartz crystals slowly grow from solution. Geology has time in abundance, and the atoms use the time to hunt for the energetically favourable ordered positions and the quartz crystal laboriously self-assembles, layer by layer. Synthetic quartz crystals of extraordinary perfection and purity are also grown from a water soup. To speed things up this is done at about 400°C, in a sealed pressure vessel. This is a commercial process which provides quartz for electronics, optics, and also as synthetic gems. Annual world production is a few thousand tons, but of such strategic value that the US Government holds several tons in the National Defense Stockpile.

The quartz crystal comes in right- and left-handed forms. Nothing like this happens in metals but arises frequently when materials are built from molecules and chemical bonds. The crystal structure of quartz has to incorporate two different atoms, silicon and oxygen, each in a repeating pattern and in the precise ratio 1:2. There is also the severe constraint imposed by the Si–O chemical bonds which require that each Si atom has four O neighbours arranged around it at the corners of a tetrahedron, every O bonded to two Si atoms. The crystal structure which quartz adopts (which of all possibilities is the one of lowest energy) is made up of triangular and hexagonal units. But within this there are buried helixes of Si and O atoms, and a helix must be either right- or left-handed. Once a quartz crystal starts to grow as right- or left-handed, its structure templates all the other helices with the same handedness. Equal numbers of right- and left-handed crystals occur in nature, but each is unambiguously one or the other.

Sand is quartz, and sand is the main raw material for glassmaking. Mixing sand with wood ash makes a glass that is workable at 900°–1,000°C. As a constituent of sandstone and as an ingredient of concrete sand is a building material of the first rank. Logically but perhaps unexpectedly, quartz sand is also the starting point in the manufacture of silicon chips. While electrical copper is 99.99 per cent pure, silicon wafers for semiconductor devices are made at 'nine nines' or 99.9999999 per cent purity (one impurity atom in a billion).

Other materials come from other rocks. In eastern Iceland, near the Reydar fjord, there is a quarry which for three centuries supplied the world with crystals of calcite. Calcite is abundant in the form of limestone, but the large and perfect crystals of Iceland spar were hugely prized, particularly by scientists and instrument makers. When the supply was interrupted in the 1880s Sir George Stokes wrote from the Royal Society in London to the Danish Government to beg for it to be restored.

Both the crystalline form and the optical properties were a perennial fascination. A stonemason was sent in 1688 by the King of Denmark to gather 'the Icelandic crystals', and soon afterwards the Danish scientist Rasmus Bartholin discovered the phenomenon of double refraction, which causes any image viewed through a calcite crystal to appear double. Thereafter Iceland spar appears repeatedly in the history of physics, in the hands of Huygens, Newton, Brewster, Fresnel, and many others. William Nicol's double prism of Iceland spar was the indispensable and universal means of producing polarized beams of light in the laboratory, used by Pasteur in his work on optical rotation in sugars, the first proof of the handedness of biological molecules. In 1925, Arthur Compton obtained a specimen of Iceland spar from the National Museum in Washington to measure with great accuracy the density of calcite. His purpose was to use this quantity to calibrate the wavelength of X-rays being employed in the new diffraction experiments. This is how we knew the size of the unit cube in gold.

Iceland spar is a connoisseur's calcite, but calcite in the form of limestone has large-scale and critical uses, in ironmaking, in cement manufacture, and in making concrete. US annual production of industrial limestone is about 1.1 billion tons, and this is perhaps 5–10 per cent of world production. The slag that is a by-product of ironmaking is a complicated mess of materials formed by reaction between components of the iron ore and lime. But because calcium silicates are among the main constituents, blast-furnace slags resemble cements, and they are now granulated and used in making concrete. Steelmaking, despite its enormous scale, produces no solid waste.

Clay is another raw material for human ingenuity. As the end result of the weathering of rocks, clays are platy crystalline particles, generally not more than one-thousandth of a millimetre wide and only a few nanometres thick. They are therefore naturally occurring nanoparticles, and have been the subject of exhaustive scientific study for at least a hundred years. A characteristic of clays is how they mix with water to make pastes that can be shaped and moulded. The flat particles cohere and slide under the potter's hand; and then in the kiln, as the water is driven out by the heat, these same flat particles slowly transform to new minerals which coalesce to form a solid ceramic material.

By combining clay and limestone, we make Portland cement. This process has the distinction of being the largest manufacturing operation on the planet, the annual production of cement (3.7 billion tons) comfortably exceeding that of steel. The science of cement is at least as complicated as the metallurgy of steel, if less well known. Like steel, success in manufacturing cement depended on getting the raw materials hot enough; and like steel, cement also arrived in the mid-19th century (Joseph Aspdin's patent is dated 1824). For cement, the limestone is decomposed easily enough in the kiln to form lime (with a large and undesirable emission of carbon dioxide), but the critical step is to drive a chemical reaction between the lime and the silicate minerals in the clay. For that, modern cement kilns work at about 1,450°C. The calcium silicate products are the active components of Portland cement: formed at high temperatures, they are thermodynamically energetic, and highly reactive towards water. Mixed with water in mortars and concretes, they react chemically to produce solid hydrates which bind together to make a strong, cohesive material. Embedding steel bars in concrete is to invent reinforced concrete, which Thaddeus Hyatt, Joseph Monier, and others did around 1877. Throughout the world, reinforcing bars for concrete are today the largest single use of steel.

String molecules

In 1928, a hundred years after cotton manufacture was industrialized, Kurt Meyer and Herman Mark at the Ludwigshaven laboratory of IG Farbenindustrie put together the evidence that cellulose is composed of simple sugar molecules strung together by plants to form long chains. Such an entity is a polymer, or a macromolecule. The existence of polymers was highly controversial at the time, and the Meyer–Mark cellulose structure was the first clear-cut instance. X-ray experiments were at the heart of the argument. It was an early step in macromolecular X-ray crystallography, the methods which led 30 years later to the Watson–Crick structure of DNA. From cellulose, the explanatory power of the polymer concept then spread fast in many directions: to underpin molecular biology; to support the chemical industry in developing plastics, rubbers, coatings, and adhesives; and to stimulate research towards new carbon materials such as the nanotube and graphene. In 1956, Hermann Staudinger, who had been the first to use the word macromolecule 35 years earlier, won the Nobel Prize for promoting one of the most pervasive ideas of 20th-century science.

We should take note of cellulose as the most abundant material in the biosphere, the main component of green plants. Polymerizing sugars (specifically beta-glucose, for there are many sugars) to make the chain achieves two things: glucose dissolves in water and cellulose does not; and cellulose is strong and tough (and glucose is not). One other thing as well: the small sugar molecule is a tiny, more or less round entity, but the cellulose molecule is thousands of times longer than it is thick, with a strong directionality. It is a linear string molecule, capable of building a fibre and then a thread. The individual polymer molecule in cotton cellulose consists of about 10,000 glucose units, each comprising five carbon atoms and one oxygen atom in a ring with some hydrogen and more oxygen atoms attached. The glucose unit is about 1 nanometre long so that the entire cellulose chain from end to end is about 10 microns (μm) in length (that is, about one-hundredth of a millimetre). It is obvious then that even a single cotton thread is a vast aggregate of cellulose macromolecules. In fact, it is a crystalline aggregate, as the individual molecules tend to pack together side by side to form fibrils of aligned molecules, each attached to its neighbours by chemical bonds. Thus the natural directionality of the linear chain and the tendency of molecules to form cohesive crystalline arrangements make cellulose a spontaneous fibre-former. Such is the case with many polymer materials, both natural and synthetic.

The cellulose molecule is rather simple as biological macromolecules go. It is built from a single repeat unit, all in a line, with no branches. It does not encode information as DNA does by using several different repeat units. It does not have the functional complexity of proteins, also the result of variety of repeat units. However, the presence of oxygen and hydrogen in the cellulose molecule ensures that as chains assemble side-by-side in the crystallite they are linked by chemical forces into a regular structure. The crystal is not only organized along the chain but the chains themselves are packed in a regular way.

Another simple biopolymer is natural rubber, obtained by draining sticky latex from the rubber tree. Like cellulose, the rubber polymer is a simple linear chain with a single type of unit (all right-handed by the way). The chain is thin and flexible, and with no rings. And rubber is a hydrocarbon. There is no oxygen, and the chemical forces acting between the chains are greatly reduced (compare solid sugar and methane gas). When raw rubber is produced from latex it shows no signs of crystallinity, instead the polymer chains are entangled like wet spaghetti. Only by extreme stretching do we force the chains to align side by side so that the weak chemical forces between chains have a chance to act. And only then does the X-ray scattering pattern show the telltale signs of emergent crystallinity.

Macrostructure

Cellulose is the string molecule not only at the heart of cotton, but of two other commodity materials: wood and paper. All three are of the greatest importance in material culture and trade, and have been for centuries. Paper, like cotton, is built from cellulose strings bundled into simple fibres. But in wood, the material structure is multiscale, and reflects the internal workings of living trees.

There are small trees and large trees, but none that grow taller than about 100 metres. This is a limit set by the difficulty of transporting water to the uppermost leaves. So in the growing tree materials are organized both in microstructure and at the macroscale to provide conduits for long-range flow while at the same time giving the tree mechanical stability. The fibres and channels that do this remain intact in the structure of wood. No material has a longer or more diverse (or more glorious) history than wood, and no material has a more complex structure on many length-scales. Its growth history is imprinted in its structure, and every year of its life as a tree encoded in its growth rings.

In the living tree, and in the harvested wood that we use as a material, there is a hierarchy of structural levels, climbing all the way from the molecular to the scale of branch and trunk. The stiff cellulose chains are bundled into fibrils, which are themselves bonded by other organic molecules to build the walls of cells; which in turn form channels for the transport of water and nutrients, the whole having the necessary mechanical properties to support its weight and to resist the loads of wind and rain. In the living tree, the structure allows also for growth and repair. There are many things to be learned from biological materials, but the most universal is that biology builds its materials at many structural levels, and rarely makes a distinction between the material and the organism. Being able to build materials with hierarchical architectures is still more or less out of reach in materials engineering. Understanding how materials spontaneously self-assemble is the biggest challenge in contemporary nanotechnology.

Metals, ceramics, and polymers?

Gold, sand, and string correspond (roughly) to the three big battalions in the science of materials. The metals are easy to distinguish and classify (they conduct electricity, and they look shiny like metals should). Polymers too can be picked out by checking for string molecules. But then we're left with the battalion of everything else. Traditional ceramics are the materials of fired clays, the silicates of brick, pot, and tile. By extension, glass may comfortably be included too, given that it also is a silicate material made by a hot process. At a stretch, cement as well, and even concrete.

But it is unconvincing to call everything that's still left a ceramic. There are semiconductors and important inorganic oxides, carbides, and nitrides, none of which have anything in common with traditional ceramics, except that they are all not-metals and not-polymers. There are minerals and ores. There are useful gases like hydrogen and liquids like water. And there is diamond. Perhaps it is a polymer, composed as it is of a network of carbon atoms. But it has more in common with the unclassifiable hard oxides than with usually soft polymers. So metal–ceramic–polymer pigeonholing has both overlaps and exclusions.

A more powerful framework for understanding the systematics of materials—and their diversity—is the Periodic Table. About two-thirds of the 92 elements are metals. The rest, grouped in the upper right, are mostly non-metals (carbon, nitrogen, chlorine, phosphorus, and about 15 others). A handful of semiconductors (silicon, germanium, arsenic, antimony, and bismuth) sit between the two. Many of the metals are rare and unfamiliar elements (hafnium, praseodymium, and lots more). There are at least half a dozen that everyone recognizes, such as iron, copper, aluminium, sodium, and uranium. The ceramics are compounds of a metal and one or more non-metals (calcium with silicon and oxygen for example). The polymers are compounds of carbon with another non-metal or several (hydrogen and oxygen mostly).

Every material we use is built from elements picked off the shelves of the Periodic Table. Every element is useful, although time and again new materials have come from old-friend elements like carbon, silicon, and iron, sometimes but not always seasoned with additions of less familiar ones. The elements are the ingredients, but a lot depends on the cooking.

Chapter 2
Close inspection

The idea that we can understand materials by looking at their internal structure in finer and finer detail goes back to the beginnings of microscopy, to Robert Hooke and Anton van Leeuwenhoek in the 17th century. This microstructural view is more than just an important idea, it is the explanatory framework at the core of materials science. Many other concepts and theories exist in materials science, but this is the framework. It says that materials are intricately constructed on many length-scales, and if we don't understand the internal structure we shall struggle to explain or to predict material behaviour. If we want to alter the behaviour to make better materials, we probably need to re-engineer the architecture inside. That may seem uncontroversial to a scientist, but it is certainly not the framework used by an archaeologist or a commodity trader or a surgeon in thinking about materials.

As each new microscopy arrives, we see at a finer scale, until with the scanning tunnelling microscope we now image the individual atoms. We can see how the atoms are arranged at the surface of a solid like rock-salt (Figure 6).

But with atomic resolution, we have reached our goal in spatial resolution; and attention turns next to resolving more and more

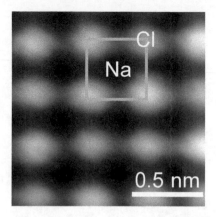

6. A direct image by scanning tunnelling microscopy of a sodium chloride (rock-salt) crystal. The chloride ions, 0.39 nanometres apart, sit on a cubic lattice, with the sodium ions between them

dynamic detail. Van Leeuwenhoek saw the beating tail of the sperm in the 16th century. Next, we want to see the atoms at play.

Down the microscope

Of course, Hooke did not see very far. His lenses were primitive, but he peered and squinted at materials such as ice, silk, cork, glass, linen, charcoal, petrified wood, and at needles and razors. He watched water rise into fine glass tubes. He described the diffusion of scarlet dye in water and the colour of peacock feathers. He looked too at limestone (Figure 7), and found that it was composed of innumerable small grains. He said they looked like the eggs of a herring. From all of this, in 1665, he made an extraordinary book, *Micrographia*. There, in his essay on flints, the shapes of crystals of sea-salt, alum, saltpetre, and ice triggered the thought that the tiny crystals he saw were all composed of regular arrangements of much smaller spheres. He did not of course see these 'globular particles', but that idea bounced around for another 250 years, until the atom was finally run to earth.

Figur: 1   *Schem. IX.*

7. Hooke's microscopic view of Kettering limestone

Crystals, crystals everywhere

There was another line of attack on the constitution of matter, that of the mineralogists and crystallographers. In the 18th century, collecting minerals was fostered by the scientific societies as part of their ambition to describe the natural world. Many minerals were obviously crystalline, so that besides their composition, interest grew in their appearance and their shape. The scientific study of mineralogy, and its sub-discipline of crystallography, began to emerge, especially in France and Germany. The most influential of the crystallographers was René Just Haüy, who carefully cleaved a fine specimen of Iceland spar to discover that the facets of the fragments all had the same angular relations. In 1822, Haüy published his huge *Traité de cristallographie* where he classified minerals by the geometry of their crystals. Later, in the middle of the 19th century, mineralogists found ways to examine thin slices of polished rock in the light microscope. Optical mineralogy, particularly making use of polarized light and chemical stains, showed the complexity of the composition and microstructure of a huge variety of rock types. It was an important event in the history of materials when in 1865 a geologist, Henry Clifton Sorby, applied similar methods to polished surfaces of metals, and revealed by chemical etching a grain structure every bit as complex as that in rocks. Metallography, using the reflected light microscope, has been an essential tool of scientific metallurgy ever since.

Then, from an unexpected quarter, in 1912, there came a great illumination. It arose from a single laboratory experiment, conceived by Max von Laue, and carried out by Walter Friedrich and Paul Knipping, three young physicists at the University of Munich. They found that a narrow beam of X-rays passing through a crystal of copper sulfate was scattered into a pattern of discrete spots on a photographic plate. Von Laue's idea was that the regular array of atoms in a crystal would act as a diffraction grating. Success would depend on a decent match between the spacing of the atoms in the crystal and the wavelength of the X-rays. Neither was known, but in the event the match was excellent. The existence of the Laue pattern proved that the crystal of copper sulfate had a regular 'spatial lattice', and patterns were quickly found also with copper metal, diamond, and rock-salt. But although each pattern was a kind of image of the lattice, it was not immediately clear how to analyse them. A simple and general way to do this was worked out by Lawrence Bragg in Cambridge a few months later. William and Lawrence Bragg then laid the foundations of modern X-ray crystallography by using these methods to solve the crystal structures of several materials, including diamond and rock salt. X-ray diffraction in the Bragg tradition is now among the most powerful of all experimental methods for solid materials. An important further step was taken in 1916 by Peter Debye and Paul Scherrer, who showed that well-formed single crystals were not essential for X-ray diffraction, but that crystalline powders and polycrystalline solids diffracted also. This hugely increased the range of materials that could be investigated.

The first few years of X-ray diffraction established once and for ever the reality of the regular lattice of atoms, ions, and molecules in most common materials. What Hooke had called globular particles were in reality atoms, ions, and molecules and, in most solid materials at least, they did occupy positions on a geometrical space-lattice which could be described by the rules of mathematical crystallography. And what Haüy had called the 'molécule intégrante' was in fact the crystallographer's unit cell. The impact of these methods on the scientific study of materials was widespread and deep. Within a decade or so, it became commonplace to know in detail the crystal structure of metals, ceramics, minerals, and the simpler inorganic and organic substances. And in the search for crystals the net was cast wide. In London and in Cambridge, J. D. Bernal investigated many unpromising, sometimes recalcitrant, materials. He started with graphite, and next found the structure of the copper–tin intermetallic compound in bell-metal bronze. He was the first to examine the hydrated silicate minerals in Portland cements: these materials were undoubtedly crystalline. But less predictably fibres of wool and silk also showed crystalline diffraction, or at least hints of it. These polymer materials and other proteins could also form crystals, improbable though it seemed, although they often needed a lot of coaxing. This profound discovery led eventually to modern macromolecular crystallography, the double helix of DNA, and detailed knowledge as well about the structure of polymer materials like cellulose. For these wonderfully intricate substances, X-ray diffraction opened the door not only to the geometry of the lattice but to the greater prize of the atom-by-atom structure of the individual molecules.

Down still further

There was impact on the science of materials also from another part of physics, as it became apparent that it should be possible to make a microscope using an electron beam rather than rays of light and glass lenses. For a while, it was not clear how exactly to do this, but in the 1930s people began to think it was a good challenge. One of them, Ladislaus Marton, wrote a little rhyme:

'A microscope,' the Walrus said, 'is what we chiefly need,
Electron Optics, furthermore, is very good indeed,
So let us build some microscopes with superhuman speed.'

In the end, two rather different types of electron microscopes were developed, one taking electrons right through an extremely thin specimen and able to resolve at the atomic scale; and another forming images from the scattering of an electron beam as it scans the surface of a specimen. The transmission electron microscope became available in the 1950s; the more versatile scanning instruments a decade or so later. The scanning electron microscope is today the most widely useful tool for investigating the internal architecture and microstructure of materials of all kinds. It has a large range of magnifications, approaches atomic resolution at its highest, and produces images of extraordinary clarity and depth. By analysing X-ray fluorescence stimulated by the electron beam it provides also detailed element maps of the image field.

In 1976, Gerd Binnig and Heinrich Rohrer built a completely different kind of microscope at the IBM Zurich Research Laboratory. Their scanning tunnelling microscope (STM) owed nothing to light or electron microscopy, and involved no rays and no lenses. At its heart was a tiny tungsten wire, with an exquisitely sharp tip, which could be brought close to a specimen surface to make a measurement of electrical current. In the STM, the tip is mounted on a piezoelectric crystal which, by applying a voltage, can be moved both up-and-down and sideways in steps as small as 1 picometre (pm), perhaps 1 per cent of the spacing between adjacent atoms. Then it becomes possible to raster the tip across the surface, and at the same time to control its height above the surface using the current that 'tunnels' between tip and surface. The image is formed from data acquired during a two-dimensional (2-D) scan over a region of the surface, either a tunnel-current map or a topographic image of the tip elevation. Atomic resolution is routinely achieved. The STM has spawned several other probe microscopies. The atomic force microscope (AFM) uses a tip which is mounted on a tiny cantilever that flexes as the tip feels mechanical interactions with the surface. The AFM has the merit of working well in liquids, and is a powerful tool for observing crystal growth and dissolution on surfaces.

Lattice work

The beauty of the space-lattice is that it describes the ultimate constitution of materials in a way that is both particular and general. When the crystal structure of diamond was first solved (by the Braggs in 1913), it became a hard fact that diamond has a cubic unit cell (Figure 8) with an edge length of 356 pm. It then follows that the distance between neighbouring carbon atoms is 154 pm (one-quarter of the edge-length multiplied by the square root of 3). Because diamond is a crystal, with a regular space-lattice, this is true for every one of the trillions of atoms in a single diamond. And it is true of the carbon atoms in every diamond that exists, anywhere, and forever. But, while these lengths and distances are unique to diamond, the geometry of the unit cell and the arrangement of atoms within it are not. Diamond shares its crystal structure with many other materials. Silicon and tin both have a space-lattice identical to that of diamond, apart only from different edge-lengths or lattice spacings (543 pm for silicon and 649 pm for tin).

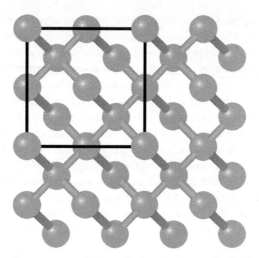

8. The cubic structure of diamond, showing the tetrahedral bonding of each carbon atom to its neighbours

The example of diamond shows two things about crystalline materials. First, anything we know about an atom and its immediate environment (neighbours, distances, angles) holds for every similar atom throughout a piece of material, however large; and second, everything we know about the unit cell (its size, its shape, and its symmetry) also applies throughout an entire crystal, in the spirit of Haüy, and by extension throughout a material made of a myriad of randomly oriented crystallites. These two general propositions provide the basis and justification for lattice theories of material behaviour which were developed from the 1920s onwards.

We know that every solid material must be held together by internal cohesive forces. If it were not, it would fly apart and turn into a gas. A simple lattice theory says that if we can work out what forces act on the atoms in one unit cell, then this should be enough to understand the cohesion of the entire crystal. It may even be a guide to what happens in a material composed of many crystallites. The particular arrangement of atoms in the unit cell is the one that gives the lowest total energy, taking into account the sum of all the forces acting between the atoms. There must be forces of attraction between at least some of the atoms, and loosely we can call these chemical bonds. In diamond, most important is the binary interaction between carbon atoms which are next to each other. Each pair of neighbours forms a strong chemical bond, the carbon–carbon (C–C) single bond. In forming this bond, the atoms are pulled towards each other, and settle at the equilibrium separation at which the attractive force between them is just balanced by a repulsive force which comes into play as they get squashed together. The equilibrium separation, the 'bond length', is the distance at which the net force acting is zero, and the potential energy is a minimum. Described like that, the chemical bond begins to sound like a spring, and so (approximately) it is. Since every carbon atom in diamond has four neighbours, it is tethered by four C–C bond-springs.

A simple way to picture the bond is as an atom-scale version of the mechanical spring that Robert Hooke first described. 'Ut tensio, sic vis': as the stretch, so the force. As we pull two carbon atoms a little distance apart, a restoring force acts. Double the distance, double the restoring force. The relation between force and distance is the same whether we push or pull. The Hookean spring is symmetrical. The potential energy varies as the square of the displacement, and it lies on a parabola with its minimum at the equilibrium rest position. Such a spring vibrates symmetrically in what is called harmonic motion.

Is such a simple model good for atoms and crystals? Well, it is and it isn't. For large displacements, the real bond-spring is quite asymmetrical. It is harder to push atoms closer together than to pull them further apart. While atoms are soft on the outside, they have harder cores, and pushed together the cores start to collide. The Universal Binding Energy Relation (UBER) shows this (Figure 9).

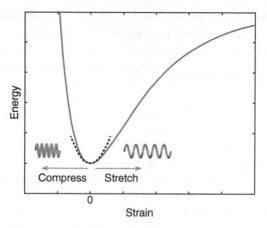

Energy

Compress ← → Stretch

0

Strain

9. The UBER curve is the solid line; the dotted line shows the energy-strain relation of the harmonic bond-spring

The UBER describes the energy-distance behaviour of a great number of materials. There is a minimum energy which defines the equilibrium bond-length. The energy shoots quickly up towards infinity as the separation is reduced and the atom cores come into hard contact, and slowly goes towards zero at large separations. But we also see that if we think only about small displacements from the equilibrium bond-length, we move just within the small cup at the bottom of the curve, and this is much the same, left or right. So for small excursions about the minimum position, the harmonic Hookean spring is a reasonable place to start.

In thinking about cohesion, the lattice does not come much into play, except that its existence allows us to think of the unit cell as a microcosm of the whole crystalline material. But there are aspects of material behaviour, especially those concerned with heat, which depend on the dynamics of the lattice as a whole. The solid material is a three-dimensional (3-D) array of atoms sitting on lattice points and connected together with Hookean bond-springs.

How does such a system vibrate? If we have two isolated atoms joined by a chemical bond, there is only one fundamental natural vibration frequency, and this depends on the mass of the atoms and the spring constant or stiffness of the bond. Now if we have three atoms joined together by two bonds, let us say C–C–C, there are now two fundamental modes of vibration, one in which two outer atoms move in the same direction, and the other where the outer atoms move in opposite directions. And in general, the more atoms that we couple together the greater is the number of modes, the number increasing in proportion to the number of atoms. In a crystal with trillions of atoms, there are trillions of vibrational modes. Of course, they are bunched close together, so we have a continuous distribution of frequencies over a range. This depends both on the fundamental spring constant of the individual bond, how the atoms are coupled together, and on the geometry of the lattice. How many are active depends on the temperature of the material. Max Born first worked out the vibration frequencies of a crystal lattice in 1912 (a good year for the lattice). For a material (such as rock-salt) with two kinds of atoms (sodium and chlorine), or for a material like diamond in which the bond-springs are at strange angles (in the carbon tetrahedron), the fundamental vibrational modes of the structure are complicated. Long after Born's theory, a heavy particle called the neutron was discovered. We can now measure the vibration spectrum of a solid material by watching how it scatters a beam of neutrons. The Born theory turns out be correct. But as the theory of lattice dynamics developed, the vibrations came to be seen as waves travelling in the solid, and having the character of particles which moved with a certain velocity, and which could collide with each other and with the atomic cores. These packets or quanta of vibrational energy are the phonons of solid state theory.

Banding together

In lattice models which describe the cohesion and dynamics of the atoms, the role of the electrons is mainly in determining the interatomic bonding and the stiffness of the bond-spring. But in many materials, and especially in metals and semiconductors, some of the electrons are free to move about within the lattice. A lattice model of electron behaviour combines a geometrical description of the lattice with a more or less mechanical view of the atomic cores, and a fully quantum theoretical description of the electrons themselves. We need only to take account of the outer electrons of the atoms, as the inner electrons are bound tightly into the cores and are not itinerant. The outer electrons are the ones that form chemical bonds, so they are also called the valence electrons. The atom of each element has a characteristic number of valence electrons (anywhere from zero to a dozen or more), depending on where the element sits in the Periodic Table. These valence electrons are quantum particles, and in an isolated atom they have a distinctive spatial arrangement around the atomic core, designed to ensure that the electrons conform to the quantum law that no two electrons shall be in the same 'state'. The state means in the same place, with the same energy and the same spin. Since electrons can have either an 'up' or a 'down' spin, this allows two electrons to be in the same place and have the same energy, provided they have opposite spins. So we can have electron pairs. In an isolated sodium atom, there is only one valence electron. But in silicon, there are four. In the lowest energy configuration, the 'ground state', two of these are paired, and the other two unpaired but spatially separated. But there are other states and energy levels available to electrons when the atom of silicon is in an excited state, for example if it absorbs a light photon.

Now when we bring a trillion atoms together to form a crystal, it is the valence electrons that are disturbed as the atoms approach each other. As the atomic cores come close to the equilibrium spacing of the crystal, the electron states of the isolated atoms morph into a set of collective states, rather like the vibrational modes of the atoms. These collective electron states have a continuous distribution of energies up to a top level, and form a

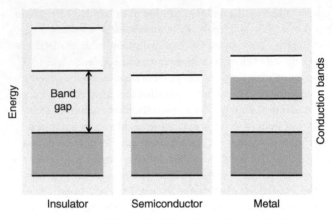

10. Band structure of different families of materials, showing band gap, with occupied energy levels shaded

'band'. But the separation of the valence electrons into distinct electron-pair states is preserved in the band structure, so that we find that the collective states available to the entire population of valence electrons in the entire crystal form a set of bands (Figure 10). Thus in silicon, there are two main bands.

They are separated by an energy gap. The size of the gap is usually measured in units of electron volts (eV). In silicon the band gap is about 1.1 eV. But of course there are also even higher bands which correspond to the excited state configurations of the isolated atom. In the ground state, the available electrons occupy all the energy states available in the band of lower energy, the valence band. The band is completely filled by electrons, which even in the crystal must obey the iron quantum rule of pairing.

This picture seems rather static, but as Felix Bloch discovered in 1928 electrons in regular periodic lattice can move freely. If the material is cold and the atomic cores have little thermal vibration, the electrons pass effortlessly as waves through the periodic lattice. As the temperature increases, the vibrations of the bond-springs perturb the motion of the electrons, which move with increasing difficulty. Alan Wilson in 1930 then realized that it is only the electrons in a partially filled band that can move freely as Bloch described. The electrons in a completely filled band, like those in the valence band of silicon, are immobile. The band gap in silicon is much larger than the thermal energy at normal temperatures, so almost no electrons reach the upper unfilled 'conduction' band. But in metals like sodium with one valence electron per atom the valence band is itself only partially filled, so the electrons are mobile. Silicon is therefore a band-gap insulator and sodium a metallic conductor.

Imperfection

The perfect crystal has atoms occupying all the positions prescribed by the geometry of its crystal lattice. But real crystalline materials fall short of perfection, sometimes slightly and sometimes spectacularly. Some imperfections concern only isolated atoms (or ions or molecules). For instance, an individual site may be unoccupied (a vacancy). Or an extra atom may be squeezed into the crystal at a position which is not a lattice position (an interstitial). An atom may fall off its lattice site, creating a vacancy and an interstitial at the same time. Sometimes a site is occupied by the wrong kind of atom. Point defects of this kind distort the crystal in their immediate neighbourhood. Vacancies free up diffusional movement, allowing atoms to hop from site to site.

Larger scale defects invariably exist too. A complete layer of atoms or unit cells may terminate abruptly within the crystal to produce a line defect (a dislocation). Or the crystal lattice may be twisted, so that what ought to be parallel planar sheets become a set of helical ramps wrapped around a line. Geoffrey Taylor and others thought that such dislocations must exist long before there was any way to see them. But in time, dislocations were observed in thin aluminium foils, using an early electron microscope, and seen to move as theory said they must.

There are materials which try their best to crystallize, but find it hard to do so. Many polymer materials are like this. In polymers such as polyethylene (PE) and polypropylene (PP), the molecular structure is regular enough, so a high degree of crystallinity is not out of the question. But as PE is cooled from the melt its entangled string molecules do not succeed completely in achieving regular crystalline order. The best they can do is to form small crystalline regions in which the molecules lie side by side over limited distances. These crystallites are embedded in disordered, amorphous regions. The individual molecules thread their way through both, and tie them together. Often the crystalline domains comprise about half the material: it is a semicrystal.

How to make a crystal

Crystals can be formed from the melt, from solution, and from the vapour. All three routes are used in industry and in the laboratory. As a rule, crystals that grow slowly are good crystals. Geological time can give wonderful results. Often, crystals are grown on a seed, a small crystal of the same material deliberately introduced into the crystallization medium. If this is a melt, the seed can gradually be pulled out, drawing behind it a long column of new crystal material. This is the Czochralski process, an important method for making semiconductors. In metal castings, crystal growth occurs along the direction of cooling, but the orientation of the crystals formed can be controlled by carefully placed seeds. Sugar is the purest everyday material manufactured as a crystal in large quantities. It is crystallized by removing water by evaporation, and adding fine seeds. Polymers are tricky to crystallize, because the string molecules easily become entangled. But by dissolving small amounts of polymer in a solvent, and giving the molecules plenty of time to find their seats, it is possible to grow faceted crystals of striking beauty (Figure 11).

For many materials, the most perfect crystals are grown from the vapour. Such were the crystals of hoar-frost forming from cold air

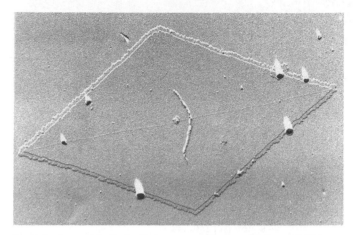

11. A single crystal of polyethylene, grown from a solution at 80°C

that Hooke observed in his microscope. And this is the route to the best synthetic diamonds.

However it is done, crystals invariably grow by adding material to the surface of a small particle to make it bigger. This may be a seed, or it may be a nucleus spontaneously formed by atom-scale fluctuations that bring together by chance a minute proto-crystal, of just a few tens or hundreds of atoms. Most of these last only for an instant, but a few survive long enough to reach a critical size beyond which they continue to grow. As new material attaches to the nucleus, the crystal starts to take shape, and identifiable crystallographic faces emerge. Different faces grow at different rates to generate the distinctive morphology of each crystalline material.

Photograph

Around 1835 William Fox Talbot put together a process for making photographic images using silver salts. It started the sequence that led from Victorian portraits to photojournalism

and the cinema, and to a multitude of images in science and
medicine (Figure 12).

For 150 years, both the still and the moving image depended on
the silver-bromide film; and then suddenly within 20 years silver
vanished, displaced by digital technology. Lawrence Slifkin, a
physicist who understood it, said that silver-bromide photography
was 'a miracle'. The slow blackening of silver salts in sunlight was
already known, but Fox Talbot discovered the *latent image*: he
found that the silver-bromide coated film holds a memory of the
image fleetingly exposed in a camera. It was invisible at first but
could be developed by chemical treatment of the film, and then
reproduced endlessly by means of a negative/positive process. The
latent image was the silver miracle.

Photographic films were improved in many ways after Fox Talbot,
but silver bromide was always at heart of them. Still, it was a
century before the latent image could be understood. It turned
out that to piece together the chain of events which connects
the incoming photon and the latent image needed a lattice model

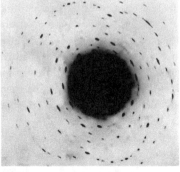

12. (a) Abraham Lincoln, a studio photograph by Alexander Gardner,
8 November 1863, from a silver negative; (b) A silver-bromide
photograph of the X-ray diffraction pattern of the beryllium ore beryl,
by W. H. and W. L. Bragg, 1915, the year of their Nobel Prize

of the silver-bromide crystal, and one comprehensive enough to include defects, chemical bonds, energy levels, diffusion, and quantum mechanics. Who was better placed than Nevill Mott, the English physicist who had worked with both Born and the younger Bragg? And so it was Mott and his collaborator Ronald Gurney who put the quantum-chemical theory of the photographic process together in a landmark paper in 1938. Here's how it works.

The film emulsion is a myriad of silver-bromide microcrystals embedded in gelatin. An individual microcrystal is several hundred nanometres (say, 1,000 atoms) across. It contains equal numbers of silver ions (positively charged) and bromide ions (negatively charged) in a cubic array, arranged so that silver ions have bromine-ion neighbours and vice versa. But there are defects, and some silver ions (which are much smaller than the bromine ions) find themselves out of position, sitting awkwardly between lattice sites. When the shutter opens, photons hit the film and strike some but not all of the microcrystals. Where a photon strikes it is absorbed in a process which detaches an electron from the bromide ion, forming at the same time an uncharged bromine atom. The electron is free to move about within the crystal. After a series of random jumps, it is trapped on the crystal surface and combines with an interstitial silver ion to form a silver atom. The whole process takes no more than a microsecond. In a microcrystal receiving many photons there is a succession of photochemical reactions like this, each depositing a silver atom on the crystal surface, and all taking place in the blink of a camera's eye.

These silver atoms aggregate into tiny clusters, and it is these which constitute the latent image. The numbers of silver atoms in a cluster may be staggeringly small, say half a dozen or so. But the silver speck is stable. Microcrystals which catch no photons have no specks. Later the film is developed by bringing an electron-donor substance (typically an organic acid, like gallic acid from tanning leather) into contact with the film. This searches out the silver specks, and by donating electrons does much what the incoming photon did originally: it converts all the remaining silver ions in the microcrystal to silver atoms, turning it black. A microcrystal which has intercepted no photons remains unchanged and unblackened. Development amplifies the latent image (hugely: perhaps 100 million times) and by making it visible forms the photographic negative. So we have an image, with a microcrystal-size pixel.

The process is extraordinary because it relies on a series of charmed steps involving photons, electrons, ions, and atoms all working together to a Goldilocks time-scale. It's a miracle that the bromine is mopped up by the gelatin before it can recombine with an electron, and that the silver atoms form a speck, and that the specks catalyse the development of the image. In 150 years, nothing else was ever found to do what silver bromide does.

In 1979, 66 per cent of US silver production went to photographic materials; by 2013 it was less than 10 per cent and falling rapidly. Silver photography is endangered if not extinct, replaced by television screens and digital cameras that trap photons in other materials and in other ways to form and store images. But would the image-rich digital world be the same if there had not been 100 years of silver photography first, and *Paris Match* and *Citizen Kane*?

Low-dimensional crystals

The ideal crystal extends indefinitely in all directions, and so it is a 3-D object with 3-D periodicity. But we can imagine an object that is a sheet, periodic in just two directions, and only one unit-cell thick. This is a 2-D crystal, and such is graphene, obtained by separating a single layer of carbon atoms from the 3-D solid graphite. Graphite is a familiar material, consisting of lattice planes of carbon atoms in hexagonal rings stacked on top of each other. It is useful as a lubricant since the sheets slide easily over each other; and an electrical conductor since electrons are delocalized in the plane of the sheet and move freely. But the discovery of single graphene sheets by André Geim and Konstantin Novoselov in 2004 was unexpected: it is a surprise that graphene is mechanically stable and doesn't rearrange itself into balls or tubes or stacks; and surprising too that it is unreactive in air and does not need protecting by a high vacuum. A sheet of graphene is a material which is all surface. In the 2-D crystal of graphene, the electrons not only bind the carbon atoms together as strongly as in diamond, but they also travel close to the speed of light as if they had no mass. And it turns out now that graphene 2-D atomic sheets are formed by the stroke of a lead pencil on a sheet of paper.

We can take another dimension out of play and imagine a one-dimensional (1-D) crystal. Commonly carbon nanotubes, nanowires and nanorods are so described. Electron transport is confined to a single direction. The band structure is radically modified. Quantum dots, which are nanocrystal clusters just a few atoms across, are 0-D materials, with electrons unable to move freely in any direction. The electrons bound in a quantum dot behave much like those in a single atom.

Non-crystals

Whisper it softly, but the world is probably a little less crystalline than materials science suggests. Consider the organic materials built around carbon: paper, textiles, wood, plastics, and rubbers. Of these rather few are crystalline. Some of the thermoplastics such as PE, PP, and POM (polyoxymethylene) are semi-crystalline, as is cellulose in paper, wood, and cotton. However, the lignin in wood, and thermosetting polymers such as the epoxy resins in adhesives and composites, are intrinsically amorphous, because they have random disordered molecular structures which it is impossible to arrange on a regular lattice. Of inorganic materials based on silicon, fired-clay ceramics, cement, and glass, and semiconductors, only the semiconductors are completely crystalline. Clay ceramics and hydrated cement contain some crystalline components, but both are predominantly amorphous or at best poorly crystalline. Glass is the archetype amorphous material, with no long-range periodicity in its structure.

Chapter 3
Tough but slippery

We agree that chalk and cheese are different. But how exactly? We can say that the chalk is hard and cheese is soft, but what precisely do we mean by hardness? Can we measure it? And what scientific meaning can we can attach to words like stiff, strong, tough... or to red; or to shiny? And turning matters around, what exactly are the qualities or properties of bronze that make it the best material for casting bells?

Light and heavy

Aerospace engineers care a lot about the density of materials. For them, lightweight is good. But sometimes heavy is good. In the oil industry drilling engineers use the dense mineral barytes to make heavy muds for controlling the pressures in the wells. And to stop the tip of a church spire blowing over in high winds, a heavy weight is sometimes suspended from it on the end of a long rod. The density appears in so many scientific equations that almost nothing can be calculated without it. For instance, it sits firmly at the centre of the Newton–Laplace formula for the speed of sound in a solid material.

As we go down the Periodic Table of elements, the atoms get heavier much more quickly than they get bigger. The mass of a single atom of uranium at the bottom of the Table is about

25 times greater than that of an atom of the lightest engineering metal, beryllium, at the top, but its radius is only 40 per cent greater. So, the density of metal uranium is 18,100 kilograms per cubic metre (kg/m^3) while that of beryllium is just 1,800 kg/m^3. The density of solid materials of every kind is fixed mainly by where the constituent atoms are in the Periodic Table. The packing arrangement in the solid has only a small influence, although the crystalline form of a substance is usually a little denser than the amorphous form: compare crystalline quartz (2,650 kg/m^3) with pure silica glass (2,200 kg/m^3). Polymers like plastics and rubbers composed of carbon, hydrogen, and oxygen have densities around 1,000 kg/m^3; silicon-based ceramics and glasses around 2,500 kg/m^3; while engineering metals such as iron, nickel, and copper lie in the range 8,000–9,000 kg/m^3. Metals of higher atomic number such as gold and platinum reach about 19,000–22,000 kg/m^3.

The range of solid densities available is therefore quite limited. At the upper end we hit an absolute barrier, with nothing denser than osmium (22,590 kg/m^3). At the lower end we have some slack, as we can make lighter materials by the trick of incorporating holes to make foams and sponges and porous materials of all kinds. We can easily get as low as 300 kg/m^3 (cork, even aerated cellular concrete). Aerogels, weird, fragile materials composed of silica membranes and which are 99 per cent nothing, have densities 20–100 kg/m^3. So in the entire catalogue of available materials there is a factor of about a thousand for ingenious people to play with, from say 20 to 20,000 kg/m^3.

Hot and cold

In about 1740, Willem 's Gravesande in Leiden built his ring-and-ball apparatus. The iron ball, made so that it would just pass through the ring, no longer does so when it is heated. His simple demonstration of thermal expansion was powerful because in metals (and most other materials) the expansion is rather small, and far from obvious without careful measurement. The 's Gravesande apparatus shows that even tiny expansions may be critical where there are tight clearances. We define the thermal expansivity as the fractional increase in length of a material when its temperature is increased by 1 kelvin (K) or 1°C . For iron, this is about one part in 100,000, so that even increasing the temperature by 100°C produces an expansion of only about 0.1 per cent. Still, in large structures, like the Forth Bridge or the Eiffel Tower, the expansions can be sizeable. The thermal expansion of course entails a concurrent increase in volume (numerically about three times the linear thermal expansion), and so the density of iron decreases by about 0.3 per cent on heating from 0° to 100°C.

Many other metals, as well as glasses and ceramics, have similar expansivities, lying say between 5–$20 \times 10^{-6}\,\mathrm{K}^{-1}$. Plastics and rubbers expand more, with values about ten times as great; and so do liquids. The expansion of materials as we increase their temperature is a universal tendency. It occurs because as we raise the temperature the thermal energy of the atoms and molecules increases correspondingly, and this fights against the cohesive forces of attraction. The mean distance of separation between atoms in the solid (or the liquid) becomes larger. In the simple lattice model of a solid, the vibrational energy stored in the interatomic bonds increases with temperature, and the vibrational amplitude gets slightly bigger. Unfortunately, the simple harmonic model of a solid does not account for thermal expansion because it has Hookean springs. The push-pull vibrations are symmetrical, so that although the amplitude increases the mean separation stays the same. The fact that we see more or less universal thermal expansion tells us that the harmonic lattice needs adjusting to reflect real material behaviour. We can build an 'anharmonic' model by making the spring weaker as the atoms move apart and stronger as they move together. There is no surprise here, because the interaction is obviously asymmetrical. For some purposes, it is a good move to neglect the asymmetry, but understanding thermal expansion is not one of them. As a general rule, the materials with small thermal expansivities are metals and ceramics with high melting temperatures. The UBER curve has a deep, narrow minimum, and adding thermal energy only increases the vibrational amplitude slightly. High thermal expansivities go with materials, like plastics, with weak cohesion, and shallow, broad UBER minima, and low melting temperatures.

Although thermal expansion is a smooth process which continues from the lowest temperatures to the melting point, it is sometimes interrupted by sudden jumps (Figure 13). We see this in iron, where there is a discontinuity in density at 912°C, as the crystal structure changes abruptly from bcc (alpha-iron) to fcc (gamma-iron). The density increase is about 1 per cent, and this means that iron contracts as it goes through the alpha–gamma transformation. Above the transformation, thermal expansion continues on its merry way. Changes in crystal structure at precise temperatures are commonplace in materials of all kinds. Quartz has a sharp crystallographic transition at 573°C, but unlike iron it expands rather than contracts, and by an enormous 5 per cent. The consequences are sometimes seen in building fires where the stress caused by the alpha–beta quartz transition causes concrete to burst explosively. A peculiarity in quartz is that above the transformation, the quartz does not show normal thermal expansion but instead contracts slightly as the temperature continues to rise. That means that beta-quartz has a negative thermal expansivity. This unusual behaviour is found in several materials, always where something occurs to mask the underlying universal tendency to expand. A venerable example is the contraction of cold water from 0°–4°C, where water molecules become more closely packed even though their individual molecular kinetic energy increases. In the case of beta-quartz, something similar occurs. Although the length of the Si–O bond increases, the arrangement of the $Si-O_4$ units in the crystal becomes more compact.

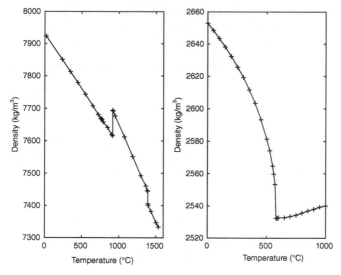

13. Thermal expansion in two common materials: how the density of iron (left) and of quartz (right) vary with temperature

In the 1890s, Charles Guillaume at the International Bureau of Weights and Measures at Sèvres discovered that iron-nickel alloys with about 36 weight per cent nickel had remarkably small thermal expansivities, around one-twentieth that of pure iron. These alloys became known as invars. They were immediately used in making all manner of precision scientific instruments. Guillaume himself ran an invar wire from the ground to a lever on the second platform of the Eiffel Tower, 115 metres above. He saw from the motion of the lever that the platform moved up and down every day by about 20 millimetres, so that 'the Tower appears as a gigantic thermometer'. Guillaume also found that the thermal expansivity could be fine-tuned by small variations of nickel content, probably the first time a material was designed to realize a precise value of an engineering property. Because they were unaffected by temperature changes, invar tapes were sent in 1899 to the expedition carrying out the geodetic survey of

47

Spitzbergen. They were so successful that invar became the material of choice in land survey work by mapmakers throughout the world. Later, for the same reason, invar was used in cathode-ray televisions to fabricate the shadow masks that form the picture dots. These two applications are largely obsolete, but it is invar now that lines the storage tanks in the giant ships which move liquified natural gas around the world at −160°C.

An extraordinary material with a thermal expansivity that is almost exactly zero is the glass-ceramic Zerodur, based on silica with additions of aluminium, phosphorus, and lithium. These ions combine with the silica to form a solid solution that stabilizes beta-quartz at everyday temperatures. Crystallites of beta-quartz, with a slightly negative expansivity, are dispersed in a glassy matrix of alumina with a normal positive expansivity. The two almost exactly cancel. The composite expansivity is at least one-thousand times smaller than that of iron. Zerodur was developed by the German glassmaker Schott as a new material to fabricate a 4-metre mirror for an astronomical telescope, the first of many as it turned out. But precision manufacturing throws up many opportunities for ambitious materials: Zerodur is now also the material used to make the gratings that align the silicon wafers for fabricating integrated circuits.

Thermal expansivity is one of the material properties that help us to answer questions about how an object behaves as it heats or cools. Besides changes of size, we may want to know how much heat is needed to produce a certain temperature rise; or how well a material transmits heat; or even at what temperatures it can be safely used (will it melt, or will it burn?). There is a cluster of properties which describe the thermal behaviour of materials. Besides the expansivity, there is the specific heat, and also the thermal conductivity. These properties show us, for example, that it takes about four times as much energy to increase the temperature of 1 kilogram of aluminium by 1°C as 1 kilogram of silver; and that good conductors of heat are usually also good conductors of electricity.

At everyday temperatures there is not a huge difference in specific heat between materials. We measure it in units of kilojoules per kilogram per kelvin (kJ/(kg K)). For metals and ceramics, it lies in the range 0.1–0.9 kJ/(kg K); for polymers just a little higher, 1–2 kJ/(kg K). But if the specific heat is rarely of much interest in engineering, it stood centre stage in the scientific development of quantum theory in the years from about 1905 onwards. The early 19th-century law of Pierre Dulong and Alexis Petit held that the specific heat per atom was the same for every element. Since 1 kilogram of aluminium contains about four times as many atoms as 1 kilogram of silver, the difference in specific heat that we just noted is explained. But not all elements conform to the Dulong–Petit law even at room temperature, and the situation gets worse as the temperature falls, when the measured heat capacities decrease greatly, and tend to zero at the absolute zero of temperature, 0 K. In his lecture at the first Solvay Conference in 1911 Einstein described a theory of specific heat in which the energy of thermal oscillations is quantized, that is, it comes in quantum parcels. The theory was substantially improved by Peter Debye who realized that what is quantized is not the vibration of individual atomic oscillators, but the phonon energy of the entire lattice.

In all crystalline materials, thermal conduction arises from the diffusion of phonons from hot to cold regions. As they travel, the phonons are subject to scattering both by collisions with other phonons, and with defects in the material. This picture explains why the thermal conductivity falls as temperature rises, and why the thermal conductivity of stainless steel is about ten times less than that of mild steel. And it also accounts for the rather low thermal conductivity of glasses and other amorphous materials, where there is no regular lattice, and therefore no true phonon modes. In fact, the energy modes in amorphous materials rather resemble Einstein's individual oscillators, and the distance between scattering events is little more than the distance to the next atom. In metals, heat energy is carried by the free electrons as well as the lattice phonons, so that the electrical and thermal conductivities are roughly proportional (the Wiedemann–Franz law). But it is single-crystal diamond (with no free electrons) that has the highest thermal conductivity of all: a testament to the easy transport of phonons in an almost perfect lattice. This makes diamond cool (to touch).

Stiff and strong

Whether materials are stiff and strong, or hard or weak, is the territory of mechanics. Our everyday language for capturing the mechanical characteristics of materials is not at all rich. To stiff, strong, hard, and weak we can add soft, brittle, tough, and slippery to make a basic vocabulary. But the scientific story of mechanical behaviour is immensely complicated, with a long history. Galileo wrote about the bending of beams, and Hooke realized that all solid materials were somehow like springs. In the 19th century the mechanical theory of how materials stretch and bend under the action of forces was described by French mathematicians, notably Siméon-Denis Poisson, Augustin-Louis Cauchy, and Gabriel Lamé. Rigorous results were obtained for an idealized Hookean elastic material, one which deforms when a force is applied, and then returns to its initial state as soon as the force is removed, and in which all deformations are proportional to the applied forces. These ideal materials were considered to have no internal microstructure, because at that time little was known, although there was much speculation. (It might have surprised Hooke that a theory that ignored the evidence of the microscope could be so serviceable.) And the 19th century continuum theory of linear elasticity is still the basis of much of modern solid mechanics.

A stiff material is one which does not deform much when a force acts on it. Stiffness is quite distinct from strength. A material may be stiff but weak, like a piece of dry spaghetti. If you pull it, it stretches only slightly (in fact, the change in length may be hard to see), but as you ramp up the force it soon breaks. To put this on a more scientific footing, so that we can compare different materials, we might devise a test in which we apply a force to stretch a bar of material and measure the increase in length. The fractional change in length is the strain; and the applied force divided by the cross-sectional area of the bar is the stress. To check that it is Hookean, we double the force and confirm that the strain has also doubled. To check that it is truly elastic, we remove the force and check that the bar returns to the same length that it started with. If that is all in order, then we calculate the ratio of the stress to the strain. This ratio is the Young's modulus of the material, a quantity which measures its stiffness. Pieter van Musschenbroek, a friend of 's Gravesande in Leiden, was the first to build an apparatus to do this.

While we are measuring the change in length of the bar, we might also see if there is a change in its width. It is not unreasonable to think that as the bar stretches it also becomes narrower. The Poisson's ratio of the material is defined as the ratio of the transverse strain to the longitudinal strain (without the minus sign). The Poisson's ratio is a quantity which varies from one material to another, and in a revealing way. We might imagine that, when we stretch the bar, we simply change its shape, so that the narrowing exactly compensates for the increase in length while the volume remains the same. It that were so, Poisson's ratio would be exactly one-half. There are a few rubbery materials with Poisson's ratio close to one-half, but in most materials it lies well below, often around one-third, and in some cases closer to one-quarter. This means that when we stretch the bar, the narrowing is too small to compensate completely for the extension of length. Surprisingly, the volume of the bar increases slightly as we pull it.

There was much argument between Cauchy and Lamé and others about whether there are two stiffness moduli or one. The argument concerned isotropic materials, those which have the same properties in all directions, but since many engineering materials are isotropic it had practical significance. In fact, there are two stiffness moduli. One describes the resistance of a material to shearing and the other to compression. The shear modulus is the stiffness in distortion, for example in twisting. It captures the resistance of a material to changes of shape, with no accompanying change of volume. The compression modulus (usually called the bulk modulus) expresses the resistance to changes of volume (but not shape). This is what occurs as a cube of material is lowered deep into the sea, and is squeezed on all faces by the water pressure. The Young's modulus turns out to be a combination of the more fundamental shear and bulk moduli, since stretching in one direction produces changes in both shape and volume.

The Young's modulus is a useful practical measure of stiffness just because it combines both fundamental modes of deformation. The stiffest material yet known is the carbon nanotube, with a Young's modulus of about 1,000 gigapascals (GPa), to the extreme right of the Ashby chart in Figure 14, and slightly stiffer than diamond.

Iron and steel are about one-fifth of this, about 210 GPa. To get some sense of the numbers, if we dropped a cube of steel to the bottom of the Mariana Trench (about 10,000 metres water depth), it would be compressed in volume by only about 0.05 per cent. A cube of polyethylene would be compressed by more than 10 per cent. A factor of about 10,000 covers the useful range of Young's modulus in engineering materials. The stiffness can be traced back to the forces acting between atoms and molecules in the solid state, as reflected in the UBER curve. Materials like diamond or tungsten with strong bonds are stiff in the bulk, while polymer materials with weak intermolecular forces have low stiffness. Both the elastic stiffness and the thermal expansivity depend on the depth and steepness of the UBER curve. It is not surprising then to find that stiff materials show low thermal expansion, and that high thermal expansion goes with

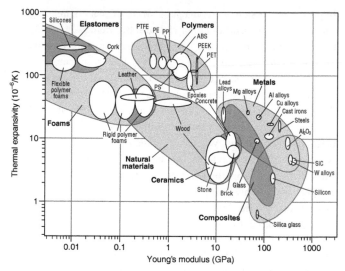

14. Ashby charts, named after Michael Ashby, engineer and materials scientist, show the relation between two selected properties for many materials. This one shows thermal expansivity and Young's modulus

low elastic modulus. The strong top-left to bottom-right trend is striking in the Ashby chart.

Thermal expansion and bulk compression are both ways of altering the density of a material, and are really part of the same story. We can combine information on both to produce a master formula, the equation of state, which tells us what the density is at any temperature and pressure. What happens deep in the Earth shows how materials behave under the most severe conditions. The Earth's core is made of solid iron at a temperature of about 6,000 K and a pressure of about 350 GPa (about 3.5 million atmospheres). The estimated density is about 13,000 kg/m³, just a little less than twice the density of unstressed iron at the Earth's surface. The opposing effects of pressure and temperature have squeezed the atoms closer to each other, and reduced the interatomic spacing by about 30 per cent.

In pure compression, the concept of 'strength' has no meaning, since the material cannot fail or rupture. But materials can and do fail in tension or in shear. To judge how strong a material is we can go back for example to the simple tension arrangement we used for measuring stiffness, but this time make it into a torture test in which the specimen is put on the rack. One of several things may happen. We find, if we test a bar of ductile brass, that we reach a strain at which the material stops being elastic and is permanently stretched. We have reached the yield point, and beyond this we have damaged the material but it has not failed. After further yielding, the bar may fail by fracture, but we may even be able to draw the bar out a long way, and this is how we make a copper wire. On the other hand, with a bar of cast iron, there comes a point where the bar breaks, noisily and without warning, and without yield. This is a failure by brittle fracture. The stress at which it breaks is the tensile strength of the material. For the ductile material, the stress at which plastic deformation starts is the tensile yield stress. Both are measures of strength.

It is in metals that yield and plasticity are of the greatest significance and value. In working components, yield provides a safety margin between small-strain elasticity and catastrophic rupture. While yield is damaging, it is rarely dangerous. But, still more important, plastic deformation is exploited in making things from metals like steel and aluminium. In metal-forming operations like strip rolling and sheet drawing the yield stress is exceeded deliberately so as to achieve a permanent reduction in thickness or change of shape.

The theory of plasticity complements that of elasticity. In 1864, the French mechanical engineer, Henri Tresca, described how metals deform when they are punched or extruded through shaped holes. He claimed that metals yield when the shear stress exceeds a critical level, different for different metals, a yield criterion which remains useful today. Later Richard von Mises proposed that yield occurs when the stored energy arising from the shear strain exceeds a critical value, a more generally accurate rule. But explaining why metals yield as easily as they do took longer. Laboratory experiments showed that large single crystals of metals like zinc deformed by slipping along certain crystallographic planes, and they do so at surprisingly low stress. The slip planes were parallel but spaced many thousands of atoms apart. The isotropic plasticity of most metals was no doubt the resultant of a multitude of slip events on randomly oriented grains. Yakov Frenkel calculated the expected yield stress of an ideal crystal lattice and found that it was much larger than observed. Clearly something occurs that lets ductile metals slip irreversibly far below the Frenkel stress. And X-ray experiments show that the lattice spacing in a plastically deformed metal is the same as in the undeformed metal, so whatever is happening has little effect on the bulk of the material. Of course that means the deformed metal retains its cohesion and its strength beyond the yield point. The explanation is that real crystals are not perfect, and plasticity has its origin in defects. Point defects cannot achieve a great deal, but an extended defect such as an edge dislocation can sweep through an entire crystal and in doing so displaces the lattice by one interatomic spacing. This produces a small overall deformation, without a restoring force to return it to where it started. So the deformation arising from dislocation movement is plastic and irreversible. To account for bulk plasticity a huge number of dislocations must sweep though the material. But not only do vast numbers of such dislocations exist, even more are created as deformation proceeds.

A useful feature of plastic deformation in metals is that plastic straining raises the yield stress, particularly at lower temperatures. Cold-working produces strain-hardening. This happens because as dislocations are multiplied during working, they tend to become entangled and pinned on features such as grain boundaries. When the stress is removed there is elastic recovery, but if a repeat test is run under tension the specimen retains its configuration of less mobile dislocations. It then takes a larger stress to initiate slip than it did in the first test. The metal can however be brought roughly back to its original state by annealing it at a high temperature.

Viewed on the large scale, plastic deformation appears entirely smooth, but in sensitive tests on microcrystals it is possible to detect sound waves from dislocation movements. These come in acoustic bursts, caused by avalanches of dislocations. Nano-plasticity is evidently erratic. Metal-forming may not be so easy on the nanoscale.

Brittle failure is not only noisy but often scary. Engineers keep well away from it. An elaborate theory of fracture mechanics has been built up to help them to avoid it, and there are tough materials to hand which do not easily crack. Several of the big ideas in fracture mechanics came from Alan Arnold Griffith, who studied brittle failure in glass at the Royal Aircraft Establishment around 1920. The noise of breaking glass reminds us that as the stress builds up in a brittle material, so does the stored elastic energy. When it breaks, the noise is a side effect of suddenly releasing that energy. Griffith argued that most materials have small cracks and flaws in them, and that in a brittle material the tip of a Griffith flaw is a sharp feature where the general stress can be hugely amplified. As the loading increases (or as the flaws get bigger), there comes a point where the largest of them grows uncontrollably. The crack feeds on the elastic energy around it, and runs rapidly through the material. Griffith showed that for a given material and a given stress, there is a critical crack length for instability. Since small cracks and flaws are present in almost any engineering component (like a train rail), the trick is not to avoid cracks but to avoid long cracks which exceed the critical length. In brittle materials, the local tip stress tears apart the atoms or molecules as the crack runs. In materials which can yield, the tip stress can be relieved by plastic deformation, and this is a potent toughening mechanism in some materials. Scientifically speaking, a tough material is one that resists failure by cracking or tearing.

Around 1660, there was great enthusiasm for Prince Rupert's drops, made by letting molten glass drip into a tank of water. These little glass tadpoles, each with a long tail, resisted breaking with a hammer or by squeezing in a vice, and were a lot of fun at the Royal Society. They are tough because as the hot glass falls into the water, the skin cools rapidly and solidifies, while the glass inside is still warm and liquid. As the interior then slowly cools it shrinks, and pulls the solid glass skin inwards and compresses it. Later, the compression of the surface makes it difficult for a crack to develop as this can only happen under the action of tension stress. (Breaking off the tail returns things to normal, as it releases the surface compression.) The trick of compressing a material to suppress cracking is a powerful way to toughen materials. Automobile windscreens are toughened in this way by chilling the surface in a blast of cold air, and Gorilla glass in mobile devices is compressed by stuffing the glass structure with large potassium ions. Concrete beams can be put into compression by pre-stressing with steel tendons.

Not all cracking is fast and violent. There are cracks everywhere, and most of them do not bother us much. Old paintings are often covered with a web of craquelure, which arises from the gradual shrinkage of the surface (Figure 15).

The varnish on oil painting hardens and shrinks as it ages. Its shrinkage is restrained by the underlying picture and it develops a tensile stress as though it were being stretched. Any tiny flaws in its surface act as stress magnifiers, and small cracks grow to relieve the stress. Each subsequent crack tends to develop roughly at right angles to the previous one, and little by little a network of intersecting cracks develops.

In describing the mechanics of materials, we have a mountain to climb, as the behaviour across all materials is wonderfully various. Unfortunately, linear elasticity takes us only to base camp. For a start, the relation between strain and stress may not be linear.

15. Vermeer, *Girl with a Pearl Earring*: map-cracking of the varnish

A more dramatic deviation is that it may not even be constant with time. We see these time dependencies clearly in polymer materials. If we place a load on a steel beam, then it bends in the middle, but the amount of bending stays forever the same. But if the beam is made of polyethylene, the deflection slowly increases with time, and if after a long time we remove the load, it only partly recovers. The beam stays bent. This phenomenon is called creep. It may also affect metals, but only if they are used not too far from their melting temperatures.

The mechanics of rubber is even odder. A rubber band can be stretched to perhaps three or four times its length, and it always returns obediently to where it started. This 'high' elasticity is nothing to do with UBER springs, which at the absolute theoretical most can only reach an elastic strain of 10 per cent (and in practice much less because of yield and fracture). And in the rubber band test, we notice that there is a definite limit, a maximum stretch, at which the material resists further extension.

In natural rubber and other elastomers, elasticity arises from the thermal agitation of the polymer molecules. A single molecule can be stretched to its full length, but when the ends are released the thermal motion of the atoms along the chain causes it to crowd up into a more disordered, and less extended, configuration, like a snake writhing in a box. If we keep an eye on the distance between the two ends of the molecule, we see this is at its greatest when the molecule is fully extended, but this distance becomes smaller as the molecule reverts to a random coil. In fact, the end-to-end distance reduces to something close to the square root of the chain length. The forces resisting the stretching of a rubber have little to do with the attractions and repulsions between atoms or the stretching of chemical bonds. They are mostly to do with the entropy of the random chain, which decreases as the chain is stretched. Molecular systems fight against any process which decreases entropy, and this appears as a force opposing stretching. This mechanism becomes more vigorous as the temperature rises, so that rubbers become stiffer as they get hotter. This contrasts sharply with the behaviour of metals, and indeed of solid thermoplastics.

Hard and slippery

Hard is different from tough. Hardness is a property which materials scientists think of in a particular and practical way. It tells us how well a material resists being damaged or deformed by a sharp object. That is useful information and it can be obtained easily. There is for example the Brinell test, named after the manager of a Swedish steelworks who in 1900 devised a simple way to check the quality of batches of steel plate. He placed a steel ball bearing between two pieces of plate and squeezed the sandwich in a vice. The ball makes a permanent circular impression, the size of which is a measure of hardness. This test and others like it have been standardized, and are widely used. For a given load, the pressure under the indenting ball can easily be calculated if the diameter of the indentation is measured with a microscope. For many metals, this pressure turns out to be proportional to the yield stress.

What about slippery? According to the classical laws of friction (Amontons' laws), the force required to achieve steady sliding of a block across a surface (of steel on ice, let us say) is proportional to the downward force acting on the block (often, just its weight); and the force does not depend either on the area of contact or on the sliding speed. The coefficient of friction is just the constant of proportionality; and it depends both on what is sliding and the surface it is sliding on. Amontons' laws work well for hard stiff materials (ceramics and many metals), but less well for soft, deformable materials such as viscoelastic polymers (thermoplastics and rubbers).

The sliding of two rough surfaces against each other is just one of many ways to dissipate mechanical energy. Often dissipation is spread throughout the body of a material rather than at its surface, for example when a material is deformed. This happens in its purest form in the viscous flow of a simple liquid, when the shearing movement is entirely converted to heat; and also during plastic deformation in metals, when most of the work done on the material as it yields also ends up as heat. In viscoelastic materials, some combination of dissipative and elastic processes occurs, and dramatically in the soft rubbers used in the tyres of Formula 1 racing cars. Here viscoelastic deformation and surface wear together absorb much mechanical energy, reducing the elastic response of the tyre, and increasing the friction with the road surface. This works best when the rubber is hot, and the natural vibrations of tyre-on-track are matched to the time-scales of molecular relaxation. The heavy price is the short lifetime of the tyre. Small amounts of dissipation may even occur within metals which are working more or less elastically below their yield point. This *internal friction* arises from slip at grain boundaries, or when dislocations and other defects move under an applied stress. This brings us back to bells, since it is the unusually low internal friction of the tin bronzes which underlies the persistence of the sound of the struck bell. A good bell also radiates a strong sound. For this we need a high speed of sound, which the Newton–Laplace formula shows requires stiff material. The bell-metal bronzes meet this requirement too. The woods with outstanding musical qualities and traditionally used to make the soundboards of violins, such as spruce and pine, also combine low internal friction, high stiffness and high sound velocity.

Soft

Soft is sometimes the opposite of hard, so that chromium is hard and chalk is soft. But a different kind of soft is squidgy. Squidgy materials were beyond the scientific pale for a long time. Pierre-Gilles de Gennes took soft matter seriously, and won a Nobel Prize in 1991. In the soft box, we find many everyday materials like foods, paints, and medicines. Some soft materials such as adhesives and lubricants are of great importance in engineering. For all of them, the model of a stiff crystal lattice provides no guidance. There is usually no crystal. The units are polymer chains, or small droplets of liquids, or small solid particles, with weak forces acting between them, and little structural organization. Structures when they exist are fragile. Soft materials deform easily when forces act on them, sometimes even under their own weight, so that they may slump like a spoonful of mayonnaise. They sit as a rule somewhere between rigid solids and simple liquids. Their mechanical behaviour is dominated by various kinds of plasticity.

You make Indian ink by collecting black soot and mixing it with water. But soot and water don't mix well, because the soot particles are rather greasy and form clumps. The trick is to add a small quantity of gum arabic, and shake well. Then the clumps disperse to give a smooth densely black ink. The gum arabic, which is harvested as the sap of the acacia tree, is a polysaccharide string molecule, which spontaneously attaches itself to the surface of the soot particles to mask the greasiness. The gum arabic acts at the surface of the soot particles to negotiate a better relationship between soot and water. It is one example of a surfactant. Because of their fragility, soft materials are hypersensitive to small quantities of substances which modify the forces acting between molecules or particles. Gum arabic, which has been used as a dispersant and stabilizer for thousands of years, is used today in many foods, drinks, paints, and still in inks. Similar dispersants are used in making concrete, so that by making it easier to mix and disperse the cement powder, smaller quantities of water are used, and the hardened concrete is stronger.

Soft materials may or may not have a yield stress, but if they do it is extremely low. Often, it is useful to know how they deform above the yield stress, and so they are often studied in rheological tests which measure how they flow when they are sheared. A device for doing this is a Couette rheometer, in which the soft material fills the gap between two concentric cylinders. One cylinder is fixed while the other rotates. The soft material impedes motion of the rotating cylinder, and from this the viscosity can be calculated.

Liquid and gas

There are many useful materials which are liquids much less complicated than ink, and others which are gases. Gasoline, for example, and helium and alcohol and water. Let's take water. To a chemical engineer water is an extraordinary solvent; while to a mechanical engineer it is the working fluid in a steam engine.

Once James Watt had made the steam engine into a practical machine, Sadi Carnot, Rudolf Clausius, and William Thomson worked out the theory. In doing so, they cleared up the complicated relation between heat and work, and invented thermodynamics. Thomson and then J. Willard Gibbs defined the properties of steam (or any working fluid) needed to design a heat engine operating in a repeating thermodynamic cycle. The definitions were precise but abstract. These properties had still to be measured. What are the numbers? How does the pressure of a mass of steam vary as it is heated or cooled? How much heat is needed to change its temperature; and how much heat is released when it condenses?

Hugh Callendar at Imperial College in London gave most of his life to that difficult, meticulous work. Early on, he invented the platinum resistance thermometer, used today wherever an accurate temperature must be measured, and a niche market for platinum. The specific heat of water was one of his greatest interests. In 1915, Callendar published the first of the many editions of his authoritative (and lucrative) *Steam Tables*. A reviewer said 'We do not think there is any case of an experimental physicist since Regnault's time [the 1840s] doing even half as much service to engineering'. Steam trains are things of the past, but steam turbines today produce four-fifths of the world's electricity.

Cryogenics is the branch of engineering which does useful things with materials which are extremely cold. It relies heavily on fluids like nitrogen and helium which can be liquified by cooling to act as refrigerants. The industrial liquifaction of air supplies both nitrogen and oxygen. Liquid nitrogen is used as an industrial coolant in vast quantities. Liquid oxygen (more hazardous) is an oxidant in rocket fuel systems; and basic-oxygen steelmaking was designed around it. Natural gas (mostly methane) is liquified at about −160°C so that it can be transported by sea. More interesting scientifically is helium. So weak are the cohesive forces acting between helium atoms that helium gas liquifies only at about 4 K (4 K above absolute zero). Heike Kamerlingh Onnes made liquid helium in 1908, and almost immediately discovered superconductivity in metals. Today, helium is indispensible as the unique coolant for all superconductive materials and devices, including magnetic resonance imaging (MRI) scanners in medicine, and innumerable scientific instruments. A hundred tons of liquid helium are needed to cool the superconducting magnets of the Large Hadron Collider.

Carbon dioxide (CO_2) is the curious case of a gas with few uses beyond fizzy drinks, but which is now a commodity thanks to carbon credits and emission trading. The EU, where the largest carbon market operates, trades about 8 billion tons of CO_2 a year. Of course, the commodity concept is turned on its head here: in order to reduce how much CO_2 is released into the atmosphere, regulators require industries to pay for the right to produce CO_2 gas by the ton and then throw it away. This arrangement provides the incentive either to avoid producing it, or else to do something useful with it. But as a useful material, CO_2 is dismal prospect: as a source of carbon, it is thermodynamically unreactive. The best large-scale non-polluting use discovered so far is for driving hydrocarbons out of oil and gas reservoirs.

Burn and bang

There are materials whose job is to provide us with energy. That we can routinely raise a passenger aircraft (say 200 tons) 10,000 metres into the air and slide it across the Atlantic is testament to the energy density of jet fuel. The fuel, which is a refined hydrocarbon liquid, burns in the turbofan engine to produce a hot gas of CO_2 and water. This chemical reaction releases about 43 megajoules (MJ) of heat energy for each kilogram (kg) of fuel. Since the hydrocarbon molecules are the same in composition as the string molecules of polyethylene (PE), the specific energy of polyethylene plastic and aviation fuel are much the same. That heat is released if PE is caught up in a building fire. The specific energy of cellulose (and hence of wood) is only 17 MJ/kg, because the cellulose polymers already have a lot of oxygen in them, so are partially pre-oxidized. The specific energy of pure hydrogen is much higher, 140 MJ/kg. When it burns in air it produces only water, but of course we have to make the hydrogen in the first place, and storing a gas is difficult. The specific energy of other hydrocarbons is much the same. Natural gas is slightly lower, 37–40 MJ/kg, because it contains a few per cent of nitrogen and CO_2 which contribute nothing to the combustion heat.

Acetylene is a little higher, because the carbon–carbon 'triple bond' packs some additional energy. Mixed with pure oxygen, it produces an intensely hot flame for welding and cutting steel.

In explosives, the aim is to have one big bang, and to achieve the greatest possible energy-release rate. Nearly all high explosives, civil and military, are organic substances with C, H, O, and N atoms, usually in ring structures. So the fuel and the oxidant are combined in the same molecule. These materials are chosen to be as stable as possible, except when deliberately detonated. Then, the explosive decomposes in milliseconds to generate CO_2, water, and nitrogen gas. The specific energy of explosives such as TNT is only one-tenth that of jet fuel, but the rate of energy release is much higher.

At the other extreme are batteries, from which we want low power for a long time, combined with the ability to recharge. Lithium-ion batteries do both these things. The internal lithium chemistry has a high energy density, not much different from jet fuel, but the specific energy of the battery calculated from the weight of the packaged battery, rather than the active ingredients inside it, seems rather low, say 1 MJ/kg.

A nuclear fission reaction releases hugely more energy than any chemical process. A uranium dioxide fuel pellet has a specific energy about 70,000 times greater than jet fuel, but it is much more difficult to liberate it.

Chapter 4
Electric blue

Thomas Edison had interests in many materials. He set up a company to make Portland cement and helped to construct the Yankee Stadium. He believed passionately that concrete was just the thing for making furniture, and even pianos.

His electrical inventions were more successful. They included the incandescent electric light, public electricity supply, sound recording, and the movie camera. All are ingenious applications of electrical phenomena in materials. What they show is that electricity can drive devices and machines to produce light, sound, and movement. Electricity is the great enabler. Many of Edison's inventions came before the discovery of the electron by J. J. Thomson in 1897, as of course did the electrical discoveries of Michael Faraday, Werner von Siemens, and Heinrich Hertz. But it was the quantum theory of electrons in solids that explained much of the electrical behaviour of materials. It underpins semiconductor technology and all its pervasive consequences.

Good conductor, bad conductor

Electrical processes are invisible, even if their consequences are not. We don't have any everyday words to describe electricity in materials. So we start at the beginning. If we can pass an electric current through a material, then it is a conductor; if we can't, it is an insulator. In the middle, there are the semiconductors. The material property we use to organize this information is the electrical resistivity (or its inverse, the conductivity). We can make similar pieces of different materials, apply a certain voltage to each in turn, and measure the current which flows. The currents are in inverse proportion to the resistivities of the materials: the greater the resistivity, the smaller the current. We find that the range is enormous. In the best insulators, the resistivity is so high as to be infinite for all practical purposes. In the strange case of superconductors it is strictly zero.

In pure metals, the resistivity is extremely low, silver being the lowest of all. But other common metals are not far behind, and a factor of ten covers all of them. The best are used for electrical connections as diverse as minute bondwires in semiconductor circuits (gold, copper, or aluminium), telephone wires (usually copper), and long-distance power transmission cables (usually aluminium). In the metals, the low resistivity (or, put another way, the high conductivity) arises from the existence of a conduction band in the solid which is only partly filled. Electrons in the conduction band are mobile and drift in an applied electric field. This is the electric current. The electrons are subject to some scattering from lattice vibrations which impedes their motion and generates an intrinsic resistance. Scattering becomes more severe as the temperature rises and the amplitude of the lattice vibrations becomes greater, so that the resistivity of metals increases with temperature. Scattering is further increased by microstructural heterogeneities, such as grain boundaries, lattice distortions, and other defects, and by phases of different composition. So alloys have appreciably higher resistivities than their pure parent metals. Adding 5 per cent nickel to iron doubles the resistivity, although the resistivities of the two pure metals are similar. We use alloy metals like nickel–chromium and iron–chromium–aluminium as heating elements for which high electrical resistance is essential. Tungsten in the filament of an incandescent lamp has such a high melting temperature (3,420°C) that it can be safely run at white heat where its resistivity is ten times greater than when cold.

Resistivity depends fundamentally on band structure. So aluminium metal, with excellent conductivity, has electrons in a partially filled conduction band; but alumina (aluminium oxide), an important industrial insulator, has electrons in a filled valence band and no accessible conduction band (put another way, it has an ionic lattice in which electrons are transferred from aluminium to oxygen; and no electrons are mobile). The resistivity of solid alumina is about one billion billion times greater than that of aluminium metal. But even alumina turns into a conductor when we dissolve it in molten cryolite as we do in producing aluminium industrially. Then the aluminium and oxygen ions are separated and themselves become mobile within the melt. The aluminium ions migrate in the electric field of the electrolytic cell, and carry the massive electrolysis current, depositing aluminium metal on the cathode of the Hall–Héroult cell. Here it is aluminium ions rather than electrons that are the charge-carriers. At 900°C, the conductivity of the melt is about 1,000 billion times greater that of solid alumina.

Plastics and rubbers also are usually insulators. Polymer molecules are constructed of covalent chemical bonds, commonly C–C, C–H, C–O. The valence electrons are firmly locked into the bonds, and are not mobile. One exception is graphite, in which valence electrons in the plane of the carbon sheets are free to move. This two-dimensional electrical conductivity is found in its purest form in graphene, which has a resistivity only slightly lower than that of silver. Electronically conducting plastics would have many uses, and some materials are now known. Like graphite, they depend on constructing polymer molecules with mobile valence electrons. Polyacetylene was the first example, where alternating single and double bonds along the carbon backbone of the molecule provide the mobile electrons. Pure polyacetylene has a resistivity in the semiconductor range. But doping polyacetylene with iodine produces a material with a hugely greater conductivity, close to that of silver or copper, and for good measure with a metallic sheen.

There are a few elements in the Periodic Table that are classified as semiconductors, in particular silicon and germanium in Group IV of the Periodic Table. As elements of the highest purity, they are really band-gap insulators, with a low conductivity which increases with rising temperature. But when they were first studied, they were not pure, and had a somewhat higher conductivity. A few binary compounds such as copper oxide and silicon carbide were also found to be semiconductors. For a long time they were of little interest, since it was not obvious what could be done with a semiconducting material. Eventually, Ferdinand Braun discovered that a simple rectifier (to convert alternating current to direct current) could be made from a semiconductor crystal such as lead sulfide in contact with the end of a metal wire. These cat's-whiskers were used as the detectors in early radios. Interest in semiconductors perked up.

In pure silicon, the band gap is so large (about 1.1 eV) that at normal temperatures only a tiny number of electrons have enough thermal energy to reach the conduction band. The concentration of intrinsic charge-carriers is extremely low. But if a few (say 1 per cent) of the silicon atoms are replaced by atoms of phosphorus, the situation changes radically. Phosphorus, a Group V element, sits next to silicon in the Periodic Table, and has one more electron. This electron is not needed to form the chemical bond in the solid structure, and occupies a new band which sits in the band gap of pure silicon, not far below the conduction band (Figure 16).

Silicon may also be doped with a Group III element such as aluminium or gallium, with one less electron. There is then a deficiency of bonding electrons. The missing electron behaves as a

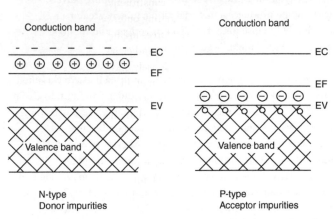

16. Bardeen's band diagram of n- and p-type semiconductors from his 1956 Nobel lecture

positive hole in the valence band, and such materials are called p-type semiconductors. Positive holes are mobile, and act as charge carriers like the excess electrons in the n-type semiconductor. These doped materials are extrinsic semiconductors in which the conductivity is directly controlled by the amount of dopant.

The billion-billion-fold difference in conductivity between metals, with large numbers of itinerant electrons which move freely through the lattice, and insulators, with localized electrons rooted to the spot, is enormous and striking. Across the entire range of materials, there is a great divide between the two groups. As we swap one element for another across the Periodic Table we seem to reach a point where materials flip abruptly between metal and insulator. In materials taken as a whole there is some kind of metal–insulator transition in play. And we discover that the simple certainty that some elements such as hydrogen are always insulators is not correct. We find that a compound such as such indium tin oxide, even though it is transparent, has free electrons and conducts like a metal…and so it is metal.

When we build a material from its constituent atoms, there is a competition between the forces holding outer (valence) electrons to individual atomic sites and the forces arising from the lattice as a whole. When atoms with weakly bound electrons (soft and fluffy, or polarizable) are inserted into a densely packed lattice their electrons tend to wander off into the metallic crowd. The electrons of less polarizable atoms in less densely packed lattices tend to be localized. This is the basis of the simple Goldhammer–Herzfeld criterion, which predicts that there is a critical density at which an insulator should become a metal. High-pressure experiments in which materials are squeezed to increase their density show just such metallization in many insulators, for example in silicon, phosphorus, and iodine, and probably in hydrogen too.

Superconductor

The electrical resistivity of many metals falls to exactly zero as they are cooled to very low temperatures. The critical temperature at which this happens varies, but for pure metallic elements it always lies below 10 K. For a few alloys, it is a little higher. Once established, an electric current in a superconducting coil continues to flow indefinitely, without dissipation, provided the current and the associated magnetic field are not too high. Of the metallic elements, niobium has the highest critical temperature, 9 K, and wire used for winding superconducting coils is usually a niobium alloy, often niobium–titanium. The alloys, particularly when work-hardened, can operate at much higher magnetic fields than pure niobium, and the wire is robust. Superconducting windings provide stable and powerful magnetic fields for magnetic resonance imaging, and many industrial and scientific uses.

It was found in mercury that specimens with different isotopic composition (and therefore different atomic mass) had different critical temperatures. These vary inversely as the square root of the isotope mass. Evidently, the electrons in a superconductor somehow feel the atomic vibrations of the lattice. The quantum theory of superconductivity in metals shows how electrons are associated in pairs through coupling to lattice phonons, and then behave quite differently from the unpaired electrons present above the critical temperature. In particular, they move freely through the material, and do not experience the scattering that is responsible for normal electrical resistance. The dissipation which usually accompanies the flow of electric current does not occur.

In 1986, Georg Bednorz and Alex Müller at IBM discovered superconductivity in a doped copper oxide or cuprate. Its critical temperature of 35 K was strikingly high. Since then many more 'high-temperature' superconductors have been found, mostly other complicated cuprates, but also some iron compounds. The record for the highest critical temperature has inched upwards, and now stands at 133 K in a mercury barium calcium cuprate (even higher values can be achieved under extreme pressure). These materials differ hugely in composition and crystal structure from the conventional metal superconductors. There is no agreed theory to explain why they are superconductors.

Magnet

A magnetic field can be created by sending an electric current through a wire coil. But it can also be produced by a permanent magnet made from a ferromagnetic material. A permanent magnet requires no power. Its magnetization has its origin in the motion of electrons in atoms and ions in the solid, but only a few materials have the favourable combination of quantum properties to give rise to useful ferromagnetism. In fact, only three pure elements (iron, cobalt, and nickel) are ferromagnets at practical temperatures. Ferromagnetism disappears completely above the so-called Curie temperature. The highest known Curie temperature, 1,117°C, is that of cobalt.

Iron, cobalt, and nickel are neighbours in the first transition series of the Periodic Table. All three have atomic structures with several electrons with parallel spins. As a result, each atom is a strong magnet. When the atoms are close to each other as they are in solids, a rule of quantum mechanics favours a state in which the atomic magnets point in the same direction as their neighbours. In an unmagnetized ferromagnet, alignment occurs within magnetic domains, but the domains are randomly oriented so there is no net magnetization. The Curie temperature is a critical point above which thermal agitation overwhelms the long-range ordering of the atomic dipoles, and domains of spontaneous magnetization no longer exist.

Below the Curie temperature, ferromagnetic alignment throughout the material can be established by imposing an external polarizing field to create a net magnetization. In this way a practical permanent magnet is made. The ideal permanent magnet has an intense magnetization (a strong field) which remains after the polarizing field is switched off. It can only be demagnetized by applying a strong polarizing field in the opposite direction: the size of this field is the coercivity of the magnet material. For a permanent magnet, it should be as high as possible. Such 'hard' high-coercivity magnetic materials can be created by adding other elements to iron, nickel or cobalt, to form either metal alloys or ceramic materials. Of the two dominant permanent magnet materials, one is of each kind. The ferrite ceramic $BaFe_{12}O_{19}$ serves the low-cost market (say one million tons a year worldwide), and the iron–neodymium alloy $Nd_2Fe_{14}B$ is used for high-performance magnets, for example in hard-disk drives, electric vehicles, and wind turbines (say 25,000 tons). Neodymium is a rare-earth element, so this alloy is called a rare-earth (RE) magnet material, although of course it is composed mainly of iron. RE magnets have a much more intense magnetization than the ferrites. Permanent magnets are ubiquitous but more or less invisible components of umpteen devices. There are a hundred or so in every home, and perhaps half as many again in every car.

There are also important uses for 'soft' magnetic materials, in devices where we want the ferromagnetism to be temporary, not permanent. Soft magnets lose their magnetization after the polarizing field is removed, as the domains re-establish themselves. They have low coercivity, approaching zero. When used in a transformer, such a soft ferromagnetic material links the input and output coils by magnetic induction. Ideally, the magnetization should reverse during every cycle of the alternating current to minimize energy losses and heating. These were severe in the earliest iron-core transformers, but Robert Hadfield (again) in Sheffield and Ernst Gumlich in Berlin found, more or less at the same time, that adding silicon to iron made the material much softer magnetically. Silicon transformer steels yielded large gains in efficiency in electric power distribution when they were first introduced in the 1920s, and they remain pre-eminent. The best soft magnets have coercivities 100 million times lower than the best hard magnets.

In ferromagnetism there are nice examples of underlying connections between different material properties. For example, the invars (which we recall are iron–nickel alloys) are ferromagnetic, and the anomalously low thermal expansivity disappears at the Curie temperature (about 230°C). This is a strong hint of a connection. In fact, magnetization produces a small spontaneous volume increase in the fcc invar crystal structure as the material adjusts its lattice spacing to maximize the quantum-mechanical exchange interaction. When the temperature is raised towards the Curie point, the magnetization decreases, and the effect of the associated magnetostriction diminishes. This produces a contraction which roughly cancels the normal thermal expansion of the lattice to produce the invar effect. But above the Curie temperature, this offsetting no longer occurs, and the thermal expansion of the invars is normal. Such magneto-volume effects are not rare. Sometimes the effect of magnetostriction is 'anti-invar', and increases the thermal expansivity; this occurs in austenite (gamma-iron), and is an undesirable property in austenitic stainless steels.

Magnetism in solids, like superconductivity, depends almost entirely on quantum mechanics for its theory. It is true that the existence of the Curie temperature can be understood from simple lattice models, but the all-important exchange interaction in ferromagnets arises completely from a quantum description.

Transparent, opaque, or shiny

It's easy to find everyday words to describe how materials look. Sight is the sense with which we engage most completely with the materials around us. A transparent material, like glass, is something we can see through, something that the light goes straight through. If it's opaque, we can't and it doesn't. Materials can be shiny and transparent (like a diamond), and they can be shiny and opaque (like polished silver). What is happening?

First, it is not quite true to say that light goes straight through a transparent material. A ray of light passing into a sheet of glass only goes straight through if it strikes the glass at right angles. Otherwise, the direction of the ray changes, because the glass refracts. The change in angle is different in different materials. If we can measure it, then from a formula of classical optics (Snell's law, but known much earlier than Snell) we can calculate the refractive index of the material. The bigger the refractive index, the greater the change of direction. For soda glass, the refractive index is about 1.4, but for diamond it is about 2.4. Pierre de Fermat (he of the Last Theorem) rightly suggested that when the light ray enters the glass, it slows down. The speed of light in the material is reduced in proportion to the refractive index. The frequency of the electromagnetic wave is unaltered, but its wavelength becomes shorter.

Another thing which happens as the light ray first strikes the transparent material is that a small fraction of it is reflected back. This is dealt with by another optics equation, from Augustin-Jean Fresnel, which describes how much is reflected and how much travels on into the material as the refracted ray. The material property that fixes this is also the refractive index. The higher it is, the larger is the fraction reflected. When the ray reaches the back surface of the sheet, the same thing happens again. Some of the light is reflected back inside the material, and the rest passes out into the air and continues on its way. The part that is reflected back then itself splits again when it hits the front surface, and so on, and so on. There are multiple reflections and refractions in 'going straight through'. If we add them all up, we find that for a ray of light striking a pane of common window glass at right angles, about 96 per cent of the light gets through and 4 per cent is reflected back (look at a window from a lighted room at night to see the back-reflection most clearly). The proportion reflected increases greatly if the light ray strikes the pane obliquely. For diamond, the amount reflected at all angles of incidence is much greater than for glass, about 30 per cent at right angles. The splitting of the light as it encounters surfaces is greater for diamond than for glass, and this underlies the brilliance of a cut diamond. The facets are cut to return as much light as possible to the observer by successive Fresnel splitting.

There is another characteristic of a gemstone diamond and that is its 'fire': the colours which appear as it is viewed in white light. A simple Newtonian prism disperses white light into a rainbow of colours because the refractive index varies with the wavelength of the light. The red ray and the blue ray in white light follow slightly different paths as they propagate through a transparent material. In a parallel-sided sheet of glass the different colours recombine to form white light when the transmitted light emerges; but if the surfaces are not parallel (as in a triangular prism), they do not recombine and the observer sees colours in the transmitted beam. Such effects occur in lenses (where they are undesirable 'chromatic aberrations', and are cancelled by ingenious optical engineering), and in gemstones where they are highly desirable, and enhanced by skilful faceting. The extent to which colours are seen in different materials depends on the difference in the refractive index to red and blue light, a quantity known as the dispersion. The dispersion of diamond is 0.045, the highest of the gemstone minerals, and about five times higher than crown glass. Dispersion in glasses is sensitive to the exact composition, and special glasses have been developed with low dispersion for lenses and other optical components. Similar low dispersions can be achieved in polymer glasses such as acrylic (PMMA).

Many simple liquids, water for instance, are colourless and transparent. Optically, they closely resemble glasses. But water becomes muddy and opaque when it is contaminated with fine particles. Mix a little clay or chalk into a jug of clear water and it no longer transmits light. The ray of light now encounters particles at every turn as it propagates into the liquid, and at every encounter there is a Fresnel splitting. The light is repeatedly scattered and in more or less random directions. Small particles with a high refractive index enhance scattering. If the thickness of liquid is great enough (or the concentration of particles high enough), no light makes it through to the other side: the material is opaque. If some diffuse light gets through, it is translucent. In solid materials, transparency is also commonly reduced or eliminated by scattering from heterogeneities. The majority of strikingly transparent materials are either non-crystalline glasses or single crystals, since both of these types of material have uniform internal structures. Polycrystalline solids have grain boundaries which produce diffuse scattering, and porous materials have internal surfaces. Gemstone sapphires are alumina single crystals and transparent, but alumina ceramics made by sintering alumina fine particles scatter light strongly and are opaque. Powdered cassiterite, a tin oxide mineral with a high refractive index, has been used to make opaque white glass since the 2nd century BCE.

The incident light may penetrate a long way into an opaque material. When it finally returns to the surface from which it entered it is scrambled and has lost all memory of the direction of the incident ray, so we speak of diffuse reflection. But by shiny materials we mean materials which are like mirrors, which reflect specularly. The requirements are that the surface should be flat, so that Snell's law gives the same angle of reflection at every point on the surface, and also that the penetration of light is extremely shallow. Polished metals fit the bill best: silvered layers on household mirrors, beryllium mirrors coated with gold in space telescopes. Penetration of light below a depth of a few nanometres is prevented by free electrons, which have perfect reflectivity over most if not all the visible spectrum.

...And red

There is a profusion of words for colours. Just for red, we have carmine, scarlet, crimson, burgundy, vermilion, ruby, magenta, and many more. There are materials (pigments, dyes) which are chosen primarily for their colour; there are many other materials which must be coloured as part of the making of things such as hats, and cars, and flags.

Pigments and dyes work by selectively removing particular colours from the spectrum of white light. The light which enters the eye is then no longer white, but consists of the white spectrum minus the parts that have been removed. The pigment vermilion, made from the rare mercury sulfide mineral cinnabar, takes out blue light, so that it appears brilliant red. Cinnabar is a semiconductor with a band gap corresponding to light of about 620 nanometres, so that it absorbs all the blue and violet parts of the spectrum. Many natural mineral pigments are semiconductors of this kind, and often contain sulfides. The element sulfur itself is a strong yellow colour because of photon absorption at the violet end of the white-light spectrum. The mineral lapis lazuli, which was the source of highly valued ultramarine pigment, is an aluminosilicate mineral in which trisulfide ions are responsible for absorbing red light, producing an intense blue. Glass has been coloured since ancient times by adding small amounts of metal oxides and sulfides to the melt. Colours arise from the absorption of parts of the white-light spectrum through transitions in the electronic structure of the metal ions. Cobalt glass is an intense blue, while dissolved uranium produces the distinctive yellow-green colour of 'vaseline' glass. Chromium glass is deep green, even though chromium dissolved in corundum produces ruby gemstones. The absorption wavelength shifts as the atomic environment of the metal ion changes.

Other pigments and dyes were obtained from plants, insects, and molluscs. Red carmine, used by painters like Tintoretto and in Chinese furniture lacquers, is from the cochineal insect. The active molecular component is an organic acid in which three fused carbon rings have an electronic structure with energy levels separated by just the right amount to allow absorption of blue light. Until the mid-19th century, all strongly coloured dyes and pigments were natural materials, usually rare and costly.

The democratization of colour occurred in a burst of discovery in a few years from the mid-1850s, when William Perkin, Heinrich Caro, and others showed that it was possible to make organic dyestuffs on an industrial scale. The first synthetic dyes were derived from coal-tar aniline, a by-product of making coke, and the first of them was mauve. Today, denim is dyed with synthetic indigo, first made from aniline too. Other families of synthetic dyestuffs also closely resemble natural materials. The anthroquinones, such as the alizarin reds used in printing and textile dyeing, are based on the same carbon-ring structure that is present in cochineal. The mid-19th century was a time of great progress in the fundamental science of organic chemistry. August Kekulé showed that carbon rings were common in molecular structures, and August Hofmann demonstrated how the molecular structure of these complicated molecules can be modified by rational sequences of chemical reactions. At the same time, there was a surge in demand for dyes as the production of textiles was industrialized. Later in the century, dyestuffs manufacture gave shape to an expanding chemical industry, particularly in Germany. Much of its chemical know-how was to form the foundation of the pharmaceuticals industry.

Not all colour comes from dyes and pigments. The red hazard light may consist of a white lamp seen through a lens, perhaps of PMMA coloured with a red anthroquinone dye, but it is more likely that it shines red because a light-emitting diode (LED) sends out red photons. The bright red light from the LED, powered by a low-voltage battery, is produced by electroluminescence in a semiconductor such as indium gallium nitride InGaN. Electrons and holes recombine to release photons at an energy fixed by the band gap. Common in nature too are colours produced by interference between light reflected from two surfaces separated by a distance less or not much more than the wavelength of light. Such thin structures and such iridescent colours are seen in pearls and peacock feathers, as Robert Hooke observed. But the red coloration of the Northern cardinal bird comes from an organic carotenoid pigment.

Virtual material

If we have a workable theory of how materials behave, let us say how they look when the light shines on them, or how they change shape when stretched, then we can write an algorithm to feed to a computer. More and more of the materials we meet and spend time with are virtual, and exist only as computer-generated images on a display or a movie screen. Algorithms for virtual materials are built into the techniques of computer animation so that the shattering of glass, or the swirl of clothing, are rendered realistically. Beneath these algorithms lie scientifically accurate representations of material behaviour.

Computational models are also used more and more in scientific work on materials. They provide a way to represent processes in which many things are happening simultaneously, and at many scales from the atomic to the bulk. The propagation of a crack, although controlled by the breaking of chemical bonds between individual atoms, may run long distances and involve billions of atoms. And near the crack tip, there are complicated stress fields, releasing energy which radiates outwards at the speed of sound. This combination of atom-scale, microscale, and bulk scale (and on different time-scales too) can be at least partially captured in a computer experiment. In fact, some features of fast cracking which can be seen in the computer experiment may be impossible to observe in the laboratory.

Sometimes, computational models can be used which sidestep all the quantum behaviour, and concentrate on the dynamics of the atoms, constrained for example by UBER-type bond-springs. Good predictions of the elastic and optical properties of polymers, glasses, and minerals can be made. More fundamental approaches to computing the structure and properties of materials must place the electrons and quantum mechanics at the heart of things. The goal is to be able to calculate the energy of a large assembly of atoms with their outer (non-core) electrons. Walter Kohn showed that, particularly for regular crystalline materials, there is a powerful way to do this, by computing the energy (and the mechanical, thermal and other properties that follow from this) from how the electrons are distributed, and then devising ways to calculate this electron density. These density-functional methods are versatile and powerful. They are used for example to explore how a small molecule like methane attaches itself to the cavity of a zeolite catalyst, or to calculate the band structure of a complicated semiconductor.

Computational methods are poised to become extremely powerful, and to open the way to things we cannot easily do in the laboratory. While laboratory instruments probe either the atomic or the bulk scale but rarely both at the same time, computational models may move easily between several different scales. And while laboratory synthesis is time-consuming, one element in a solid-state lattice model may be swapped for another with ease, making it possible to identify, say, complicated alloy compositions with unusual properties.

Chapter 5
Making stuff and making things

First catch your hare. Then make your pie. This two-step approach to making useful things is what we mean when we say that a ship is made of steel, or a dress is made of silk. Someone makes the steel; someone else buys it, and builds a ship. The maker has choices. A dress can be made of cotton or linen or silk or polyester, or even of paper. The two steps create a division of labour and skills between, say, the ironmaster and the shipwright; the weaver and the garment-maker. Or the quarryman and the mason. This separation leads to the trading of commodity materials and to international markets in steel, textiles, copper, and cement. And gold, ivory, and pearls.

But sometimes making the material and making the artefact merge into one. It was always true that the potter makes the pot-stuff and the pot-thing at the same time. And there are so many examples of the convergence of material and artefact in modern manufacturing that it feels like a trend. It's the way biology usually makes things. But, for glass, most metals, most plastics, cement, and much else, making the stuff and making the thing are still separate activities.

Glass

Glass has been traded since antiquity. Ingots of coloured glass have been found in a Mediterranean shipwreck of the Late Bronze

Age. Microanalysis of ancient glass tells us that at that time primary glass production occurred at only a handful of sites. Glass-blowing, invented in Italy in the 1st century CE, was always a highly skilled craft distinct from that of glass-making. And medieval Venice, then the greatest glass-working centre, imported raw glass from Syria and the Levant.

In recent decades, two spectacular innovations in glass-forming have opened up new uses for glass, but making the primary material and making the product from it are still discernibly separate. The first innovation, around 1957, was the Pilkington float process which made flat glass widely available throughout the world. A layer of molten glass floats on the hot surface of a tank of liquid tin, and is drawn off as a continuous ribbon of glass sheet, free of distortion and with a smooth surface. Before the float process, glass of such quality could only be made by laborious and costly grinding and polishing of plate glass. The easy availability of window glass (and special sheet glasses that reduce solar heating) has strongly influenced the design of buildings of all kinds. The skyscraper depends not only on structural steel and concrete, but on sheet glass (and elevators).

The second innovation was making glass fibre for digital communication. Molten glass forms fibres easily, and glass fibres are used widely for strengthening plastics and cements. For cements, alkali-resistant glasses containing zirconium have been developed. But for optical data transmission, the manufacturing tour-de-force was of a different order. In 1966, Charles Kao first showed how a silica-glass fibre can be used as an optical waveguide in which pulses of red light from a gallium arsenide (GaAs) laser are channelled along the fibre by total internal reflection. The stumbling block at the beginning was absorption loss, which could be traced to iron and other metal ion impurities in the glass, and which meant that the signal dropped to almost nothing over distances of just a few metres. Kao's opinion was that if the glass could be made pure enough it would be possible to send signals over kilometre distances. So it proved. By reducing metal-ion impurities to less than one part per billion, the loss was reduced by a factor of many thousands. More absorption was traced to minute amounts of water in the silica, and this too was removed in manufacture. High-performance silica fibres can now transmit near-infrared and visible light with minimal attenuation over thousands of kilometres. The complex manufacturing sequence to achieve this product still separates the first step of making the material from the final forming or shaping step. Making the primary glass pre-form of precise and graded chemical composition is technically distinct (and usually commercially separate) from the final process of fibre-drawing. It is done by chemical vapour deposition at high temperature, and is a supreme example of technical glass-making.

Silicon

Kao observed with delight that optical fibres for superfast communication are made from sand 'centuries old'. He might have said the same of the whole of silicon-based semiconductor technology. And like optical fibres, silicon for semiconductors depends on methods for making primary materials of extreme purity. Pure elemental silicon is made from silica sand (silicon dioxide), just as glass is. The first step is to remove the oxygen. This is done by reaction with carbon in an electric arc furnace. The product is about 97 per cent pure, and most of it goes to make aluminium–silicon casting alloys. But a little of this metallurgical-grade silicon is used as the starter material for making ultrapure silicon for electronics. By reaction with hydrogen chloride the silicon is converted to trichlorosilane, which is vaporized to leave behind most of the impurities. The trichlorosilane is then reduced once more to solid silicon by reaction with hydrogen. In a final step, the solid silicon is melted at over 1,400°C, and large single crystals of silicon are slowly formed over hours or days by Czochralski-pulling with a crystallographically aligned seed. As the seed is oriented, so grows the entire crystal. The single-crystal ingot, typically 300 millimetres in diameter and up to 2 metres long, is sliced into wafers 0.5–0.8 millimetres thick, and precision-polished. These wafers are then traded into the semiconductor industry.

Superalloy

Nickel–aluminium superalloys are the materials that make air travel by jet liner possible. They are used to make the turbine blades in the hottest part of the engine. The blades spin with a tip speed of nearly 500 metres per second, while the clearance between the tips of the blades and the housing is about 0.4 millimetres, the thickness of four sheets of paper. The temperature of the streaming combustion gas in which they work is about 1,600°C. This is 250°C higher than the melting point of the superalloy itself, so each blade is cooled by forcing air through internal passages and out across its surface, and by an insulating zirconia-ceramic coating.

The nickel-based superalloys used today are the result of 60 years of stepwise improvement in properties and performance (Figure 17).

Nickel alloys generally work well at high temperatures. Nickel–chromium alloys have been used as heating elements for a long time. The chromium resists oxidation. Nickel alloyed with aluminium has low creep at high temperatures, and was chosen for early gas turbines. Titanium can replace part or all of the aluminium. Little by little, the mechanical benefits of adding other alloy elements, notably rhenium, have been identified. A high-temperature superalloy may now contain eight to ten elements besides nickel. All of the alloys have a two-phase microstructure in which microscopic blocks of intermetallic Ni_3Al or Ni_3Ti are embedded in a matrix of fcc nickel.

For turbine blades, the alloys are half the story, but the other half lies in how the blades are made. The first blades were formed by

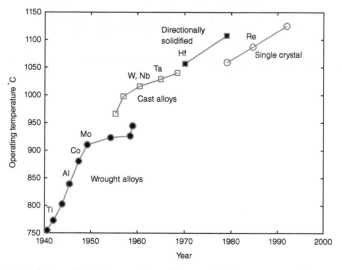

17. Evolution of nickel superalloys. The operating temperature rises as fabrication methods and alloy composition change

extrusion and forging, but later they were made hollow and lighter by casting. Next, a way was found to make stronger blades by cooling the mould at one end, to start the solidification at the base of the blade. This orients the weak grain boundaries in the direction of the centrifugal stress. Even better is to eliminate the grain boundaries entirely. Most blades are now made this way. The trick is to seed the solidification through an orifice small enough to let only one favoured crystallite grow. The nickel matrix of the whole blade is then one single crystal, and the embedded intermetallic particles are aligned crystallographically too.

String and textile

It is perhaps an odd thing that by twisting together short fibres it is possible to spin a continuous yarn. On that slender and improbable fact hangs the whole of rope-making, net-making, spinning, and the weaving of cloth (Figure 18).

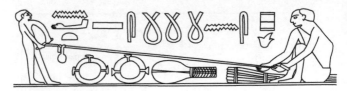

18. Rope-laying in ancient Egypt

Short-fibre yarns have strength because of the friction between individual fibres, which may be only 20 millimetres or so long as they are in cotton. This friction is greatly increased by twisting. Some natural fibres like wool have scaly surfaces which do not slide easily against each other. In cotton, the fibres are naturally crimped and interlock. But most of important of all, the twisting ensures that when the yarn is stretched the fibres press against each other to bring into play a frictional force to resist sliding. And on the microscale, the same principles can be put to work to make twisted yarns of carbon nanotubes, each of which is only 10 nanometre in diameter and about 100 microns (μm) in length. Helical twisting of the strands is also the secret of good rope-laying. The strands themselves consist of twisted yarns. In a 'classical' rope of three strands, in which the twists of the whole rope and of the individual strands are opposed, and the twist angle is ideal, the rope has maximum stiffness. This relies on the geometry of the rope rather than on the material properties of the fibres.

Strings and ropes lead us to knots. The crossings of the yarns in textile manufacture are simple loops, but true knots are formed in making nets. The knotting of yarns is found in the material culture of every society, in carpet-making, rock-climbing, sailing, surgery, and fishing. Is it possible to make knots from individual molecules? Knotted DNA molecules have been spotted in the electron microscope, and simple entanglements play a part in rubber elasticity. But true knots can now be deliberately created in smaller molecules by chemical reactions, at least in a few cases. This is an important proof-of-principle for materials, because it shows that molecules can be permanently joined together not only by chemical bonds but by mechanical interlinking. Catenanes are simple linked molecular rings; and in a rotaxane a ring-molecule shuttles up and down a rod-molecule. A pentafoil knot, with five crossings and no free ends, has been tied in a 160-atom molecule. It happens to be purple.

Plastic

At least 50 families of plastics are produced commercially today. Their names are a nightmare, so I use the standard acronyms, and give a list at the end of the book. Several are familiar materials such polyethylene (PE), polypropylene (PP), the polyamide nylons (PA), and poly(ethylene terephthalate) (PET), the bottle plastic and the polyester textile fibre. These materials all consist of linear string molecules, most with simple carbon backbones, a few with carbon–oxygen backbones, like polyoxymethylene (POM). PE, the first of these, was discovered, unexpectedly, around 1938, and others came in a rush in the decades up to about 1980. It is unlikely that there will be any more of them, because the permutations on the simple chemical structures have been largely exhausted. The early PEs were made by brute-force high-pressure synthesis. Then in the mid 1950s Karl Ziegler discovered a low-pressure route using titanium catalysts. Almost immediately, Giulio Natta found that the hydrocarbon gas propylene also polymerizes under mild conditions if Ziegler catalysts are used. All this marked a huge advance. Making PP was a triumph. High-pressure PP was a sticky mess, but the catalyst adds the propylene molecules one by one to the polymer chain so that they all face the same way. Behold, an excellent solid thermoplastic. Within only three years PP was brought to market by the Italian chemical company Montecatini. It was followed by a variety of modified PEs and synthetic rubbers from other chemical companies far and wide. Besides the half a dozen or so commodity polymers, there are many minor polymers with valuable uses in engineering. One such is polyetheretherketone (PEEK), a stiff and tough thermoplastic with a melting temperature around 350°C, high for a polymer. Drilling and reaming PEEK is much like machining aluminium. It is used in hot, aggressive places, like oil wells. Plastics as a group are valuable because they are lightweight and work well in wet environments, and don't go rusty. They are mostly unaffected by acids and salts. But they burn, and they don't much like sunlight as the ultraviolet light can break the polymer backbone. Most commercial plastics are mixed with substances which make it harder for them to catch fire and which filter out the ultraviolet light.

Above all, plastics are used because they can be formed and shaped so easily. The string molecule itself is held together by strong chemical bonds and is resilient, but the forces between the molecules are weak. So plastics melt at low temperatures to produce rather viscous liquids in which the molecules slide past each other. And with modest heat and a little pressure, they can be injected into moulds to produce articles of almost any shape, or they can be extruded through simple dies to form tubes or fibres or profiles. This transformability was seen by some as remarkable. Roland Barthes said of 'plastic' that it was a miraculous substance, but equally 'the first magical substance that consents to be prosaic'. For all their practical value, plastics had to fight long and hard to establish an aesthetic, and to find an identity as more than poor substitutes for traditional materials. Because they were protean, plastics seemed to lack inherent character. No-one railed against them more persistently than Norman Mailer, for whom 'creeping plasticism' was taking over the world. Still, at about the same time, Ben Nicholson and Pablo Picasso were happy to paint with Ripolin, a plastic house paint. Now we are addicted to plastics. Producing primary thermoplastics is a large-scale but routine business, which has largely migrated out of Western Europe and North America to China and the Middle East. There industrial plants produce plastic granules and pellets in quantity to sell to processors worldwide, who mould and extrude consumer products of spectacular variety.

Fired-clay ceramic

On the other hand, making pots doesn't work like that. It's not a two-step business. The French potter Taxile Doat wrote: 'Whether a painter, a sculptor or an architect, the ceramic artist must also be a chemist'. Nothing could make clearer how in pottery the making of the material and of the artefact are merged into a single act. And this is true of bricks and tiles as much as of Sèvres porcelain. The chemistry lies in the mineral reactions which transform soft matter into hard in the hot kiln. The shaped clay first loses the small amount of free water which makes it plastic under the potter's hand, and shrinks. As the heating gets stronger, somewhere around 400°C, the clay particles themselves begin to change. Hydroxide ions in the aluminosilicate layers of the tiny clay particles combine together to form water molecules which break free and diffuse into the kiln atmosphere. Little by little, the particles lose their crystallinity and become disordered meta-clays. As the temperature continues to rise, above about 700°C, the meta-clays in turn break down, forming a little molten glass. At the same time, new crystalline minerals, stable at high-temperature, start to appear. If the clay was rich in lime, there are calcium silicate minerals such as wollastonite. In high-fired clay ceramics, where the kiln temperature reaches 1,200°C or higher, the aluminium silicon oxide mullite can be seen in X-ray diffraction patterns. The glass, often similar in composition to volcanic glasses like obsidian, binds the complicated mixture of different particles together to form a sintered mass. It usually contains voids, so that the fired product is a porous, elastic, brittle, hard solid. The soft paste of clay particles has been converted to a coherent continuous solid material.

The character of the fired material is the result of what happens on the kiln. This depends on the mineral composition of the raw clay, on the temperature and duration of firing, and on the chemistry of the kiln atmosphere. The unfired material invariably contains not only clays (and there are many distinct clay minerals), but also quartz, perhaps lime, perhaps iron oxides, perhaps other non-clay silicate minerals such as feldspars, perhaps deliberately added pigments, and carbon tempers like straw. This rich mix of ingredients when cooked in the kiln yields a fired material equally rich in composition. But it is the variety of textures and properties which is remarkable in fired-clay materials, ranging from low-fired opaque earthenwares through stronger and harder stonewares to high-fired translucent porcelains.

Material-as-device

In 1952, the engineer Geoffrey Dummer said that it should be possible to make 'electronic equipment in the form of a solid block with no connecting wires'. It was just an idea, but the real thing was put together a few years later by Jack Kilby, who made the first integrated circuit. In the transistor, the solid-state laser, the light-emitting diode, the CCD image sensor, and a host of other semiconductor devices we see the most elaborate and extraordinary examples of material-as-device. In making any or all of them, the device takes shape as the material itself is formed, and the distinction between the material and the thing made begins to disappear. This was so in the earliest transistors, where the n-p-n junctions were formed by pulling a crystal from a melt of germanium doped with arsenic. Midway during the pull, a pellet of p-type dopant such as gallium was added to the melt, and then finally more arsenic to revert to n-type. It only remained to slice the crystal up to produce the transistors.

Kilby's first integrated circuit (IC) was rather simple, but did combine a handful of resistors, capacitors, and transistors in a solid piece of germanium. Today ICs pack hundreds of millions of transistors into a few square millimetres of material. Fabricating such devices (and other heterostructures) is done by laying down layers of material of precisely controlled composition, into which are etched the 2-D structures of the device. Performance may depend incorporating exotic elements such as hafnium.

As many as 15 layers are built up in making an indium gallium nitride blue LED, all by chemical vapour deposition, in which the elements needed for the layers are delivered in high speed gas streams flowing across the surface on which the new layer is to be formed. An initial n-type GaN layer about 1 µm thick is deposited on a substrate of sapphire or silicon nitride. The quantum wells which emit the LED light are formed by several layers of InGaN only 2–3 nanometres thick and separated by 10-nanometre layers of GaN. The colour of the light can be altered by tweaking the composition of the quantum well, and its thickness.

Laying down atomically thin layers of material is a powerful general method of constructing material/device structures. Graphene is a single-layer 2-D material but it is not the only one, so that stacked structures of graphene interleaved with other things become possible. Boron nitride is a graphene analogue, and there are 2-D sulfides like that of molybdenum, MoS_2. Building up a layer-cake superlattice is a different way to achieve material-as-device heterostructures.

Device-as-material

When we see the limitations in the properties of materials, it is natural to wonder about pushing the bounds. For example, can we have materials with a negative thermal expansivity that contract as they get hotter? In fact, we have already seen that beta-quartz does just this, and many unexceptional materials, such as calcite, do so along at least one crystallographic direction.

What about a material with a negative refractive index? Such a metamaterial bends light the wrong way. The stick in the pond not only appears bent, but bent to the wrong side. And a simple slab of such a material acts as a perfect lens, bringing diverging beams of light to a focus. By extension, a screen of such a material can steer light around an object to hide it: the invisibility cloak realized. But how to make it? That's more difficult. It requires that the two electromagnetic material properties (the electric permittivity and the magnetic permeability), which usually have positive values, should both be negative at the frequencies of interest. It seems unlikely that we shall find such a material unless we do some microengineering to achieve it. So we are talking about making a device which is designed deliberately to act as a material with a negative refractive index: in other words, device-as-material.

While it is not obvious that a negative refractive index can be found in a solid material, it can be engineered with an array of small resonators. Since this only works if the resonators are smaller than the wavelength of the radiation, the proof-of-concept was done with microwaves. Here we see again the distinction between device and material dissolving before our eyes. If we can fabricate something which has within it an array of microscopic resonators and which behaves as a negative refractor, do we have a material or a device?

The concept of a metamaterial has spread well beyond its origins in negative refraction. Cloaking and invisibility for example arise also in acoustics. But more generally, the idea of a device-as-material is a fertile one. In mechanics, there has been a drive to find materials with negative Poisson's ratio, materials that get wider and expand as they are stretched. Some examples are known (they are called auxetics), but in most cases they deform in this way because of a microengineered structure of articulated ribs which move outwards when they are pulled. These too are really metamaterials.

From ingredients

There is another way to make things: industrial chemists call it formulation; cooks call it cooking. It consists in mixing ingredients together to make a useful (and sometimes succulent) substance. It may be a paint, or a medicine, or a mayonnaise. If a paint, the recipe includes a pigment to provide colour, a polymer binder to hold it all together, perhaps a filler...The science of formulation is now rather refined, because given a larder of ingredients the permutations are enormous.

It is not an easy matter to make a smooth paint from half a litre of water and a kilogram of titanium dioxide with a surface area the size of a football pitch. It requires special mixing equipment, and a pinch of something to help the water spread over the particle surfaces, and a pinch of something else to prevent the particles from clumping together (shelf-life). Add to this things to prevent it going mouldy and to make it work nicely on the brush (but not to drip). And make it all safe and inexpensive. If the paint is white, the pigment is titanium dioxide, the ubiquitous white material of the consumer world. Titanium dioxide works as a brilliant pigment because it has a high refractive index, thanks to the heavy atom of titanium. This ensures that when a pigment particle is embedded in a polymer binder in a dry paint film that light impinging on the surface is scattered here there and everywhere as the rays are strongly bent at the pigment/binder interface. This Fresnel scattering is the mechanism of opacity, and allows the paint to hide whatever it is covering.

A simple vinaigrette is made by whisking vinegar into olive oil, with salt and lemon juice. The vinegar (which is mainly water) breaks up into tiny droplets which are dispersed in the oil. But the droplets are not stable, and over hours they recombine. Eventually, the vinegar settles into a separate layer beneath the oil. Mayonnaise is another oil-and-vinegar emulsion, but this time it is the oil which forms droplets which are dispersed throughout the vinegar. Adding the egg yolks stabilizes the oil droplets. Lecithin molecules from the egg accumulate at the surface of the oil droplets, and increase its affinity for the vinegar. The oil droplets also obstruct the flow of the usually free-flowing vinegar to produce a thick viscous mayonnaise. The opaque creamy appearance is caused by Fresnel scattering of light at the multitude of oil/vinegar interfaces.

Material-as-object

Once in a while there are materials which have value as-found. The wild pearl is such an object, a thing both of beauty and of scientific interest. Calcium carbonate pops up one more time, but here not as calcite but as the mineral aragonite, a polymorph with a different crystal structure. In pearl, aragonite grows as minute platy crystallites 5–10 µm wide, but only about 0.7 µm thick. These assemble in layers, and these layers in turn are held together by sheets of organic material, a polysaccharide (similar to cellulose) and a protein (similar to a component of silk). The organic layers that bind the aragonite crystallites together are extremely thin, about 80 nanometres, much less than the wavelengths of visible light. Thus they nicely produce the interference colours which give pearls their silver or gold iridescence as light is reflected by scattering from the interfaces. It is not likely that oysters take much pleasure in these beautiful colour effects, but the same microstructure of oriented aragonite crystallites bonded by spacer layers of organic polymer bring mechanical advantages. The composite is strong and resists impact because small movements of plates in the polymer matrix dissipate energy, and blunt and deflect sharp cracks. In this, the oyster has a much greater interest.

The use scientific

Materials are materials because they are used. Scientists are particularly good at finding new uses for materials, as Guillaume did with invar. Here are three more. They show how science depends on making new things, and how the ingenuity lies in matching the material to the novelty.

Annual world production of the metal osmium is only a few kilograms, but the scientific value is immense. Around 1885, the

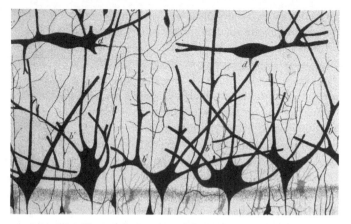

19. One of Camillo Golgi's drawings of brain tissue stained with osmium–silver

founder of neuroscience Camillo Golgi discovered that osmium (used as the compound osmium tetroxide) made it possible to bind silver efficiently to slices of brain tissue (Figure 19).

Featureless grey matter suddenly appeared full of rich microscopic structure. The histologist Santiago Cajal was enchanted: 'What an unexpected sight! Sparse, smooth and thin black filaments or thorny, triangular... black cells could be seen against a perfectly translucent yellow background! One might almost liken the images to Chinese ink drawings on transparent Japanese paper.' Osmium, the rarest of the six platinum-group metals, is now used to stain brain tissue in electron microscopy at the highest magnification. Osmium works for this because it acts chemically on the fatty molecules of the tissue, depositing osmium metal particles that strongly absorb electrons.

Almost as obscure as osmium, caesium is the least abundant of the alkali metals. It has few uses, none at all as a structural metal, but it has one supremely valuable application in the atomic clocks on which the Global Positioning System is built. More than

50 satellites in orbit carry little tubes of caesium. Atomic clocks depend on a highly accurate measurement of a spectroscopic property of an atom. In caesium, the energy level of the solitary valence electron is split by spin-coupling to the atomic nucleus (the hyperfine interaction). The gap between the split levels can be measured with a microwave spectrometer to provide a fundamental frequency standard. The time unit, the second, is now defined as 'the duration of 9192631770 periods of the radiation corresponding to the transition between...two hyperfine levels of...the caesium-133 atom'. The caesium atom is the metronome which fixes universal time. There is hyperfine splitting in the atomic spectra of many elements, but caesium is best because it has a single stable isotope (and so has the simplest hyperfine spectrum). The spectral lines are sharper in a vapour than in a solid, and the low melting temperature that makes caesium metal otherwise useless makes it easy to vaporize (caesium would melt in your hand, but your hand would catch fire). Second-generation clocks use atomic fountains in which caesium atoms are cooled close to absolute zero temperature and then puffed by laser pulses up through a microwave cavity before falling back again under gravity. The cold atoms and the long measurement interval between the up and down passes through the cavity increase the accuracy of the clock a thousandfold. Physicist Steven Chu's thumbnail is that a clock of this kind started at the birth of the universe would be off by less than four minutes today, about 15 billion years later.

If osmium and caesium are unfamiliar elements, a third example of scientific ingenuity in using materials just needs diamonds. The diamond-anvil cell (DAC), invented in 1958, allows us to go to the centre of the earth on the kitchen table (or almost). US customs officers from time to time confiscated diamonds from smugglers crossing the border. Sometimes they passed them along to other government agencies who could use them. Alvin Van Valkenberg and Charles Weir at the US National Bureau of Standards had the idea of building a DAC, and a deal was done. DACs, now found in laboratories around the world, are small compact machines you can hold in one hand. They provide a simple way to exert pressures of up to three million atmospheres (and edging higher). In a DAC, there are two faceted diamond crystals between which can be placed a speck of material, a tiny fraction of a cubic millimetre in volume. By a stiletto-heel effect, a small force exerted on the diamonds is amplified to an enormous pressure on the minute sample squeezed between the two faces of the anvil.

Why diamond? Because it is outstandingly stiff and strong, and transparent not only to visible light but also to the X-rays often used by scientists. In fact, the DAC is the usual device to carry out Bragg-diffraction experiments to look at the effects of pressure on materials of all kinds, including geomaterials such as rocks. If we shine a laser on the sample at the same time we can also heat it. In this way, we can see how rocks melt and deform at pressures found at the bottom of the mantle, 3,000 kilometres below the surface of the Earth.

Putting it all together

Most of the things we use are made from many materials, so that there is a further stage of fabrication as we put the bits together. The d.light S300 is a solar lantern for people who live without electricity. Its main components are a silicon photovoltaic cell, a rechargeable battery, and a white LED lamp. The solar panel and the lamp are packaged in polymer housings of injection-moulded ABS polymer, the Lego plastic. The lamp has a polycarbonate lens, also injection-moulded, and a woven PET carrying strap. There are internal electrical connectors, switches, a printed circuit board, and a socket of nickel-plated phosphor bronze for charging a mobile phone. The elements carbon and silicon dominate the overall material composition, but other parts of the Periodic Table are well represented. Hydrogen combines with carbon in all the thermoplastics, and there is nitrogen in the ABS, and oxygen in the PET. Silicon is the main element in the solar cell, but it wouldn't work without the help of a little aluminium and phosphorus, and minute amounts of titanium and palladium. A sheet of glass (more silicon and oxygen) covers the photovoltaic wafers. Gallium, indium, and nitrogen are at the heart of the LED; there are rare earths yttrium and cerium in the phosphor that makes the light white; and lithium in the battery. There are copper wires with vinyl insulation (containing chlorine), some silver solder and a dab of silicone-rubber adhesive. In the battery electrolyte, there is more oxygen, more phosphorus, and chemically combined iron. The only metallic iron is in 12 stainless steel screws which hold the lantern together. Stainless steel means there is a little chromium too. All in all, that is about a quarter of the elements in the Periodic Table. There are a million examples of products like this.

Pure imagination

There are many useful, surprising, beautiful, sometimes weird, and sometimes terrible things that we do with materials. But there are also things we can't do for want of the right to stuff to do them with.

It might strike a Martian as odd that we have constructed a world that is made predominantly of metals. The public, external, non-domestic world of technology is in the metallic tradition. Engines, tools, guns, and nails. But the Martian might notice that our biological world does not use solid metals at all. From the nose to the tail of a tiger, there is not a single copper rivet, or steel truss, or titanium plate. In all the elaborate chemistry of biology, there is no metabolic process that converts metallic ions to solid metals. We do not know if it was ever tried, but if it was it did not stay the journey. Biology comes down a small/soft route, in which the structural materials bearing the largest stresses are polymer fibres in tension (like hamstrings), or ceramic-polymer composites (like bone, or shell, or tooth). Biology did not find it advantageous to use the wheel either.

It is not clear (to me) if a migrating herd of wildebeest in the Serengeti or motor traffic on Interstate 87 is the better way of transporting a million individuals 300 miles from A to B. Perhaps, in a while, we shall see legged vehicles running along the highway, sometimes jostling for a lane change, and sensitive to the intelligence of the herd. And the sky may be thick with drones, like migrating butterflies. But this daydream makes a materials scientist feel inadequate because, for all our ingenuity, we do not have any materials that are able to move and act and respond like the wildebeest's leg, or even like the cilium of Leeuwenhoek's sperm. Imagining materials can go in a thousand directions, but one is towards motility. Motility is the biologist's word for being able to move spontaneously or autonomously. Materials able to move!

The motile biomaterial puts together in single functional unit a mechanism which contracts and extends, a chemical energy source to power it, and all the sensory machinery to activate and control it. This is material-as-device of a high order. Changes in the shape and size of materials in response to external stimuli are found in many soft materials, for example in the smart hydrogels developed by Toyoichi Tanaka in the 1970s. These soft materials shrink or swell rapidly when exposed to chemical triggers or laser light. But such bulk behaviour lacks the molecular precision of, say, a synthetic molecular motor. We do not yet know how to assemble molecular motors to work together in a bulk material. Nor to engineer the internal power supply. Nor to add the molecular response system. It is engineering the mesoscale between the molecule and the bulk material that we find difficult. Still it seems likely that one day from the melée of research in nanomaterials, molecular design, and synthetic biology such an autonomous micromachine will somehow emerge.

Chapter 6
Such quantities of sand

Among things that alarm us are running out of energy and making too much carbon dioxide. The materials sector contributes to both anxieties. Even if these problems went away, we would still worry about running out of materials. And we know that both making and using materials often damage our health and our environment. All these concerns come together in the search for sustainability. Using less stuff, using it for longer, and using it again are among the important ideas that are in play.

Running short

In 1914 it was widely believed that the German war effort would soon collapse for want of imported manganese to make steel. In the event German steel production increased year on year till 1918 through workarounds with available ores not used until then. In 1942, the USA faced a critical shortage of natural rubber when the supply from Malaya and Sumatra was cut (Henry Ford's hare-brained adventure in cultivating rubber in the Amazon had long since failed). Yet, within one year a synthetic rubber industry was created out of nowhere. In 1940, not a single pound of synthetic rubber was produced, but by 1943, the figure was 700,000 tons.

Later, in 1972, a small group of thinkers, the Club of Rome, published its report *Limits to Growth*. This broke new ground in using mathematical models to analyse the future of a finite world. It set out a downbeat Malthusian view of a world hitting the buffers in food, population, and resources, at least in the absence of sustained adjustments in technology. In a list of materials approaching depletion, lead was one. But many things that we made from lead in the 1970s have disappeared for reasons of public health: water pipes, paints, solders, and fuel additives have all gone. There remains only the lead-acid car battery, which now accounts for 85 per cent of world lead consumption. Lead is not in short supply today. Its inflation-corrected commodity price (a good proxy for resource scarcity) was lower in 2011 than in 1979.

There are of course pinch-points in material resources, a few perhaps intractable. They are highlighted in numerous risk lists, one from the British Geological Survey, another from the European Commission. In the red zone are materials (or elements) which are produced or found abundantly in only a few places, and with critical uses not easily substituted. The rare-earth elements appear near the top. They (and especially neodymium and samarium) are used in high-performance magnet materials for hard-disk and DVD drives, smartphones, wind turbines, and much more. Today nearly all the supply comes from China. However, there is no fundamental global shortage of REEs: mines in the USA that closed in the 1970s are reopening, there are deposits in Greenland, and new marine discoveries in Japan.

Helium gas is another material often flagged as running out. The world uses about 30,000 tons a year, much of it as liquid helium for keeping superconducting magnets cold. For cryogenic helium there is no substitute. As a breathing gas for deep-sea divers in the oil industry, there is nothing else for work below about 300 metres water depth. The helium supply story is strange. The gas is so light that any helium in the atmosphere heads off rapidly into outer space. There is so little in the air that the element was first identified in the spectrum of sunlight. Terrestrial helium comes from the radioactive decay of uranium in rocks, and accumulates in natural-gas reservoirs, from where it is obtained as a gas-industry by-product. The United States Congress took the view in 1925 that helium was a scarce strategic material for rigid airships, passed the Helium Conservation Act, and set up in Texas a vast underground helium store with a collection pipeline. All of which still exists, bizarrely given the brief history of the airship. For 85 years, the US Bureau of Land Management has dominated the world helium supply. Releasing helium to run down the reserve has given natural-gas producers little economic reason to invest in trapping it. But there are substantial helium reserves in a few natural-gas fields around the world.

These and many other similar stories show that markets and technology have been successful in finding solutions to problems of supply, even urgent ones. Still, like coal and oil, many materials, and almost all the metals, are produced from enriched fossil sources, formed by geological processes over geological time. Resources are clearly finite, and high-grade reserves are used first. But if the consumer will pay, then leaner ores can be worked. In the 19th century, Welsh smelters demanded ores with 10 per cent copper content. Today, the gargantuan copper mines of Chile and Peru work with 1 per cent ores.

The gradual shift towards lower-grade resources is an inexorable trend, but its time-scale and severity varies from material to material. There is an unresolved tension between those who consider that running short of critical resources is a strong near-term constraint on how we use materials, and those who take the cornucopian view of the economist Julian Simon that new technology always compensates for resource depletion. Biologist Paul Ehrlich predicted in 1968 that population growth would rapidly create shortages of materials and drive up prices. In the event, prices of copper, chromium, nickel, tin, and tungsten fell during the 1980s. This settled the famous Simon–Ehrlich wager in Simon's favour, but the bet did not settle the larger argument. It remains an open question whether resource scarcity will soon impose peaks in the production of important materials similar to that we expect to see shortly in oil.

Global flow

The facts about materials consumption needed to inform policy options and technical solutions are now becoming available. Analyses by Julian Allwood and Jonathan Cullen, Thomas Graedel, David Menzie, and others show how the main commodity materials flow through the economy, from reserves to production, to use, to scrap, and to recycle.

Aluminium, the most used non-ferrous metal, is produced from bauxite ores by Hall–Héroult electrolytic smelting. World bauxite production is about 250 million tons (Mt) a year. Reserves are large, but even if they weren't, aluminium could be produced from abundant clay. Bauxite is shipped internationally: the four leading producers are Australia, China, Indonesia, and Brazil. Nearly all bauxite is converted to aluminium oxide (alumina) by the Bayer process, and most of this is used to make about 40 Mt of new primary aluminium in 200 smelters located in some 40 countries. The global industry is highly consolidated, with more than half the total world production in the hands of nine companies. Primary aluminium is combined with about 30 Mt of scrap aluminium to yield 70 Mt of aluminium ingot for conversion into products for transportation, construction and innumerable other industrial and consumer uses (Figure 20).

About 20 per cent is used in manufacturing road vehicles. Aluminium comprises about 80 per cent of the weight of most aircraft, but they are only made in small numbers. About one hundred times as much aluminium goes into drinks cans as goes into Boeing and Airbus planes combined.

The most enormous global flows are in iron and steel. Blast furnaces producing pig iron for steelmaking operate continuously,

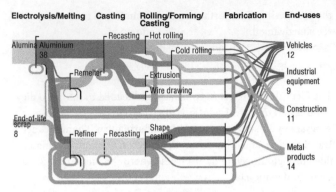

20. A Sankey diagram mapping the global flow of aluminium from production, through casting, forming, and fabrication processes, to end-uses. Numbers are million tons per year

day in, day out. A single furnace at Caofeidian in China produces 4–5 Mt of iron a year (the entire Roman Empire, which had many uses for iron, produced perhaps 90,000 tons a year). Annual world production of pig iron is now about 1,100 Mt, which combined with scrap re-use makes about 1,500 Mt of steel. To support this huge enterprise about 3,000 Mt of iron ore are mined each year, mainly in China, Australia, Brazil, and India. When at the beginning of the 20th century Homestead near Pittsburgh was the largest steelworks in the world, iron ore (a few Mt a year) went there by boat and rail from the Mesabi range in Minnesota and Ontario, about 900 miles. Now iron ore is shipped in prodigious quantities around the world. About 120 Mt moves each year from north-west Australia through Port Hedland and Dampier to the many steelworks in north-east China, including that at Caofeidian.

Cutting costs

Mining ores, making materials, and making things from them all consume energy, and directly or indirectly produce carbon dioxide. These are costs which we want to minimize. The energy used and the carbon emitted to make each ton of material or product can be calculated, though we must be careful to be clear what is included. For a material, we add up all the energy used in mining, transporting, and processing the ore, in smelting, and in primary rolling or casting. This quantity is called the embodied energy (Figure 21). It is not strictly a thermodynamic quantity, but it is a useful metric for a material, as it tells us how energy-intensive it is to make. Similarly, the embodied carbon puts a number on how much carbon dioxide is emitted during manufacture. The embodied energy and the embodied carbon are closely connected but are not the same. Any embodied energy that comes from fossil fuels contributes also to the embodied carbon because burning the fuels itself produces carbon emissions. But there may be other direct contributions to embodied carbon from manufacture itself: for cement, CO_2 is released in burning limestone in the kiln; in aluminium production, CO_2 is produced from the carbon anode of the high-temperature electrolytic cell. So for cement, the

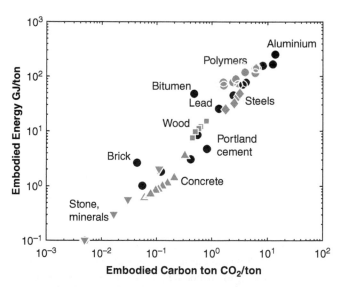

21. Embodied energy–embodied carbon for construction materials

embodied energy and embodied carbon are 5 gigajoules per ton (GJ/ton) and 1 ton of CO_2/ton; and for aluminium, 170 GJ/ton and 10 tons of CO_2/ton. One GJ is the same as 278 kilowatt hours (kWh), or roughly three weeks' electricity consumption in the average UK household.

The notion that the energy use and carbon emissions are somehow attached to materials and products for life is useful in accounting, and particularly revealing for things that are traded across national borders. A product made in one country by an inefficient industry and exported to another carries its embodied carbon with it. For example, the European Union is a large net importer of embodied carbon in aluminium.

Once the global flows are quantified, we can calculate the global energy and carbon costs. The commodity materials make large contributions to the totals because of the huge scale of their manufacture. Steelmaking produces about 9 per cent of total CO_2 from worldwide energy and industrial activities; and primary aluminium a further 1 per cent. Since making primary aluminium is extremely energy-intensive and has high embodied energy, it takes about 3.5 per cent of world electricity, some from fossil coal, some from renewable hydro and nuclear. In the round, the big five energy hogs and carbon emitters are cement, steel, paper, aluminium, and plastics.

Fit for purpose

Hadfield's manganese-steel heavy-duty rail installed in Philadelphia in 1895 lasted for 12 years in the crossings where its carbon-steel predecessor was replaced every few months. Claes Oldenburg's artwork *Soft Typewriter* (Figure 22), made of vinyl, acrylic, nylon cord, and kapok, clearly can't type letters, and is bizarre just because of its unfitness. Chocolate teapots likewise. Fuzzy as the notion is, finding fit-for-purpose materials brings a general and visible benefit, and a technology ratchet moves us that way. Some

22. Claes Oldenburg, *Soft Typewriter*, 1963, private collection

disasters are known to every engineer: the Tay Bridge, the Liberty ships (and Constance Tipper's work on brittle fracture), the De Havilland Comet jetliners, and the Space Shuttle *Challenger* (and Richard Feynman's o-ring). Things made of fit-for-purpose materials last longer than those that aren't.

If making materials and making things draws the technical eye, using things and living with them is mainly how we experience materials. Using things, or just being with them, only requires the haziest knowledge of what they are made of, and how they are put together. Not knowing doesn't stand in the way of using a cup or a car or a kaftan. While we may have little idea about the embodied energy or embodied carbon in artefacts which we buy and use, we do endow them with a value or worth. Their embodied worth may have nothing to do with market value, or with technical complexity or utility or functionality. A material object with a personal history, something hand-made, or received as a gift, is imbued with heightened value, and this may strongly affect how long it is kept. It may encourage us to mend it when it breaks. Complicated machines (refrigerators, computers) are often replaced unsentimentally and scrapped, while simple objects may be treasured for a lifetime. In ancient Rome, oil lamps made of pot had a 'use-life' of just a few months, while large jars for storing wine and grain lasted for 25 years. Decorated plates, bowls, and cups for the table were highly valued and so were frequently repaired and reused. How long we use things today, and what drives us to throw them away, are not so well known. We have a kaleidoscope of facts but little theory. Each year the average individual American throws away just over 30 kilograms of clothing, and Americans together buy about 20 billion new garments. In Japan, the average use-life of a private car increased from about eight years in 1980 to about 12 years in 2007. In Belgium, the lifespan of an electric drill is 13 years.

Using for longer is part of using less, but using less also means less material in design and manufacture or doing differently. In the First World War, Carnegie Steel supplied 20,000 tons of steel book-wire for binding US government documents, information now easily accommodated on a data stick. Lightweighting in product design means not just using lower density materials, but also removing unneeded material, often requiring some ingenuity in fabrication.

Round and round

Re-use and recycling are important targets in all views of sustainability.

An exemplary case of recycling is of the lead metal in the lead-acid battery. The situation is favourable: batteries are swapped out through an organized network of retailers, electrodes are simple in design, and the lead is never used up. Its high density makes separation easy: in the wash-tank the lead sinks as everything else floats away. In the USA, more than 95 per cent of automotive-battery lead is recycled. Recycling platinum/palladium catalytic converters from cars and trucks is similarly favourable. Repeated, more or less indefinite re-use is possible. The same is true of cryogenic liquid helium, where the slow boil-off of helium gas can be captured and re-liquified.

A more complicated but also conservative case is rhenium, a rare metal with a critical use as an ingredient of the superalloys at the hot end of an aero engine. World production of rhenium is about 50 tons a year, all of it obtained from molybdenum concentrates, which in turn are a by-product of copper smelting. To recover rhenium metal from superalloy scrap involves a lot of chemistry, and is much harder than recycling lead from batteries. But it can be done, and at $5,000 per kilogram for rhenium metal pellet it is done.

There are non-conservative, dissipative uses which entail unavoidable material loss, and then recycling and re-use are impossible: zinc coatings applied to steel in hot-dip galvanizing for instance. Much of the zinc is eventually lost by corrosion, and that which isn't is an insignificant component of the steel scrap. The extreme case of dissipative use is in burning hydrocarbon fuels. It is the use that is dissipative, not the material. We cannot re-cycle aviation fuel, but we can recycle plastics materials made from the same oil and gas.

The big wins are in recycling and re-using commodity materials like aluminium, steel, and plastics. Producing primary aluminium uses so much energy that remelting aluminium scrap brings large savings, perhaps 90–95 per cent in energy consumption. But there are some severe technical constraints. Most aluminium is used as alloys, with additions of elements such as copper, silicon, magnesium, and manganese, often in total amounts of 10–20 per cent. Once these alloy elements are mixed in to the aluminium, they cannot be removed. The usual approach to removing impurities from metals is to oxidize them and skim them off as oxide slag. But aluminium is itself easily oxidized, so this method doesn't work. Drinks cans (which are aluminium alloys) can be recycled into new drinks cans. But general end-of-life aluminium scrap, with a cocktail of alloy elements, is hard to clean up. It can be mixed with new primary aluminium alloy to 'sweeten' it; but much of it cascades irreversibly down to low-grade castings where the alloy composition doesn't matter much. There is a one-way downhill deterioration in aluminium during a recycle. Even so, about one-third of aluminium throughput is made up of recycled material, a combination of scrap from fabrication and end-of-life material.

Steel recycling is easier because it is easier to remove alloy elements. About 600 Mt of steel is recycled each year, the largest part scrap from discarded products at the end of their life. But polymer materials present a new set of difficulties. The main plastics are distinct materials which cannot be combined because polymers rarely form homogeneous mixtures that are themselves useful materials. Recycling only works if the individual plastics can be sorted into individual scrap streams. Worldwide a good deal of this is done manually, by waste pickers in India for example. Automated systems are used elsewhere, but there is a residual loss of quality compared with the primary 'new' material, and much recycled plastics scrap also cascades down to low-value uses like floor coverings and packaging. Recycling of PVC is complicated both by its chlorine content and its high content of fillers, plasticizers, and stabilizers, some of which are toxic heavy metals like cadmium. Rubber is impossible to re-use. The rubber polymers are irreversibly cross-linked by vulcanizing, so scrap cannot be remoulded. Depolymerizing it produces a material with heavy contamination by sulfur. A lot of scrap rubber ends up as filler in road building, and some is burned as a fuel in making cement. Polyethylene and poly(ethylene terephthalate) (PET) (mainly bottles) have the highest recycle rates, but overall plastics re-use is no more than 5–10 per cent, much lower than for metals.

The idea that industrialized societies are cumulatively transferring materials from their native state into a re-usable urban stockpile goes back to Jane Jacobs ('cities will become huge, rich and diverse mines of raw materials'). It may now make as much sense to meet our requirements from the urban mine as to dig in virgin ground. It is already true that the concentration of gold in a scrapheap of PCs and mobile phones (say 300 grams of gold per ton) is higher than in the richest currently worked gold mine. But the urban mine is a mixed blessing. De-mixing is always a thermodynamically uphill task. The parting of gold and silver in ancient Anatolia required salt and heat; the parting of gold and computer junk is just as difficult. To separate steel reinforcement from concrete is still uneconomic. But the size of the urban mine is too large to ignore. Over the entire 20th century about 170 Mt of copper were produced and converted into manufactured goods in North America, of which about 60 per cent remains in use, and 40 per cent has been discarded, much into landfill. This total is about 50 times the annual consumption of new copper in North America. Estimates of global stocks are full of gaps, but the global in-use copper stock is about 330 Mt. For iron and steel the global in-use stock is about 2.2 tons per capita, and for silver perhaps 100 grams. These averages hide large regional variations. The per-capita steel stock in-use is at least 10 tons in the USA, but only 1.5 tons in China.

Dust to dust

The downward cascade of high purity to adulterated materials in recycling is a kind of entropy effect: unmixing is thermodynamically hard work. But there is an energy-driven problem too. Most materials are thermodynamically unstable (or metastable) in their working environments and tend to revert to the substances from which they were made. This is well-known in the case of metals, and is the usual meaning of *corrosion*. The metals are more stable when combined with oxygen than uncombined. So the fact that aluminium metal needs so much energy to free it from its oxide ore tells us that aluminium is correspondingly unstable in contact with oxygen in the air (this is the up and down of the same process). A quirk of nature saves us (for a while): fresh aluminium materials instantly acquire a skin of aluminium oxide which is so tight on the surface that it protects the underlying metal against further reaction. But it is a fragile defence: scratch the surface and skin reforms; in seawater the salt destroys the oxide skin; and at a high temperature the skin burns off. In the thermite process, the intense heat produced by the rapid oxidation of aluminium is used for welding steel rails. Win some, lose some. By much the same mechanism, stainless steels are protected by a surface film of chromium oxide. Stainless steels also work less well in chloride-rich environments, and in the absence of oxygen to repair the film.

For plastics, the string molecules are unzipped by oxygen and ultraviolet light, generally producing small organic molecules which dissolve in water and diffuse into the environment. For some plastics, we encourage such alteration and call the materials degradable. Often it is a shortcoming and we add stabilizers to soak up the oxygen or to otherwise frustrate the degradation chemistry. But it is a holding operation, and thermodynamics always wins in the end.

Broadly speaking, ceramic materials are more stable thermodynamically, since they already contain much oxygen in chemical combination. Even so, ceramics used in the open usually fall victim to some environmental predator. Often it is water that causes damage. Water steals sodium and potassium from glass surfaces by slow leaching. The surface shrinks and cracks, so that the glass loses its transparency. Old historic glass is vulnerable since it was made with lots of sodium so it melted at a low temperature. But durable modern glass suffers too, after many cycles in the dishwasher. Stones and bricks may succumb to the stresses of repeated freezing when wet; limestones decay also by the chemical action of sulfur and nitrogen gases in polluted rainwater. Even buried archaeological pots slowly react with water in a remorseless process similar to that of rock weathering.

Collateral damage

The best estimates of the amount of copper and bronze produced in the ancient world come from chemical analysis of Greenland ice cores. Even the modest operations of the Roman smelters spewed material thousands of miles through the air. Today, there are high levels of platinum in the dust near highways, dumped as microscopic debris from the catalytic converters in every vehicle. When the damage to health from pervasive mercury pollution was recognized in the 1960s mercury production in the USA and Europe was greatly reduced. But this did not happen everywhere, and mercury is still much used in small-scale mining for gold in South America, contaminating water and food over large areas. Carcinogens like asbestos, or toxic chemicals like dioxins, which have been widely dispersed for many decades, are difficult to get back in the bottle. Radioactive waste may be hard to keep under control if, as at Hanford, America's worst contaminated site, the bottles are leaky 70-year-old concrete tanks. There are many such population-wide risks from pollution by hazardous materials.

Materials industries have often been dangerous places for those who work in them, often under inhuman conditions. Ten thousand slaves worked the silver mines of ancient Athens at Laurium. The transatlantic trades in cotton and slaves were entangled for three centuries. In the 19th century unspeakable atrocities were committed in the Belgian Congo in the pursuit of wild rubber; later, to work on the rubber plantations of Malaya was little better. In the Monowitz concentration camp, where Primo Levi was a prisoner, new kinds of atrocity were perpetrated in the ruthless effort to manufacture synthetic rubber. Today, valuable metals such as tantalum make their way into consumer electronics along murky supply chains which may start in conflict zones such as the Democratic Republic of Congo. And the jade and ruby mines of Myanmar run on forced labour and heroin.

Other categories of damage are in hazards to the environment and to wildlife. Industrial open-cast mining despoils the landscape; refining bauxite to make aluminium creates lakes of caustic, lifeless red mud; and illegal logging for wood and paper contributes to rampant deforestation. The trade in ivory rests on the persecution of elephants.

No conclusion

Sustainable materials is not really something to have easy conclusions about. Sustainability is more a frame of mind, or a checklist to see how we are doing. No doubt the list will get longer, but today it is roughly this. We are comprehensively and universally dependent on material artefacts. This is true of all societies, diverse though they are. The scale of global material flows is enormous, and now greater by an order of magnitude (or several) than for all of human history. In a finite world, there are limits somewhere, and so it is wise, on some time-scale, to favour a circular economy of closed systems (repair, re-use, recycle) over open systems (extract, make, use, discard). Whatever else we do, it helps if we reduce the rate of flow of materials through the economy (use less, use longer).

The big commodity materials (let's say steel, cement, plastics) are the industrial offspring of coal, coke, oil, and gas. Their global flows depend on abundant energy, of which they are voracious and majority consumers. These materials industries are important cogs in the worldwide shift to non-fossil fuels. There is also the rising trend of global carbon dioxide. As top energy users, commodity materials are unavoidably in the thick of it, and some of them are big direct carbon emitters as well.

Using up materials resources, taking a big slice of world energy, and emitting a lot of carbon dioxide are three features of the global materials system which need fixing, and almost certainly in reverse order of urgency, with reducing carbon dioxide at the top. But also on the sustainability checklist are: the well-being of people working throughout the materials supply chain (a lot of people), and people living with things made from materials (everyone); and the well-being of the physical and biological environment (everywhere).

We built a world of steel in a century of cheap energy, and before we cared much about carbon emissions. As we now tackle the challenges of energy and carbon, we bear down on steel. So central is steel in every society, and so coupled are all materials flows, that the consequences must be felt everywhere. Hold the thought.

Appendix 1
A very short guide to quantities and units

Multiples

milli (m), one-thousandth
micro (μ), one-millionth
nano (n), one-thousand-millionth (one-billionth)
pico (p), one-million-millionth (one-trillionth)

kilo (k), one thousand
mega (M), one million
giga (G), one thousand million (one billion)

Length

1 millimetre (mm) is one-thousandth of a metre (m)
1 micron (μm) is one-thousandth of a millimetre

1 nanometre (nm) is one-thousandth of a micron
1 picometre (pm) is one-thousandth of a nanometre

The wavelength of blue light is about 450 nm
The wavelength of red light is about 700 nm

Temperature

The melting point of ice is 0 degrees Celsius (°C), or 273.15 kelvin (K)
Absolute zero of temperature is −273.15°C, or 0 K
The melting temperature of silver is 962°C

Force, pressure, and stress

A small apple of mass 0.1 kilogram (kg) sitting on a table exerts a force on the table of about 1 newton (N)

The water pressure at a depth of 10 m is about 98,100 N/m², or 98,100 pascals (Pa). This is about 1 atmosphere (1 atm) pressure

Energy and power

To lift the 1 N apple to a height of 1 m requires energy of 1 joule (J)

To lift the 1 N apple to a height of 1 m in a time of 1 second (s) requires a power of 1 watt (W)

A kitchen kettle, rated at 3 kilowatts (kW) power, and which boils 1 kg of water in 2 minutes, uses 360 kilojoules (kJ), or 0.1 kilowatt hour (kWh), of energy

If the supply voltage for the kettle is 240 volts (V), then the electric current is 12.5 amperes (A)

The electron volt (eV) is often used as a unit of energy for processes involving electrons and photons. A photon of blue light has an energy of 2.8 eV, and a photon of red light an energy of 1.8 eV. A thermal energy of 1 eV corresponds to a temperature of 11,600 K.

Appendix 2
Abbreviations for some common polymers

ABS	acrylonitrile-butadiene-styrene	
EP	epoxy	
NR	natural rubber	
PA	polyamide	*nylon*
PAN	polyacrylonitrile	
PE	polyethylene	*polythene*
PEEK	polyetheretherketone	
PET	poly(ethylene terephthalate)	*polyester*
PMMA	poly(methyl methacrylate)	*acrylic*
POM	polyoxymethylene	*acetal*
PP	polypropylene	
PS	polystyrene	
PTFE	polytetrafluorethylene	*teflon*
PUR	polyurethane	
PVC	poly(vinyl chloride)	*vinyl*
SBR	styrene-butadiene rubber	

Appendix 3
Nobel Prizes for materials

Over the last 100 years, about 15 per cent of the Nobel Prizes awarded for physics and chemistry have been for topics closely related to materials.

1905	Adolf von Baeyer: dyestuffs
1908	Gabriel Lippmann: colour photography
1913	Heike Kamerlingh Onnes: liquid helium
1914	Max von Laue: X-ray diffraction
1915	William Henry Bragg, William Lawrence Bragg: X-ray diffraction and crystal structure
1918	Fritz Haber: ammonia synthesis
1920	Charles Guillaume: invars
1954	Linus Pauling: chemical bond
1956	Hermann Staudinger: macromolecules
1956	John Bardeen, Walter Brattain, William Shockley: transistor
1963	Karl Ziegler, Giulio Natta: polymer synthesis
1964	Charles Townes, Nikolay Basov, Aleksandr Prokhorov: laser
1972	John Bardeen, Leon Cooper, Robert Schrieffer: superconductivity
1974	Paul Flory: macromolecules
1977	Philip Anderson, Nevill Mott, John van Vleck: magnetic and disordered materials
1982	Kenneth Wilson: critical phenomena
1986	Ernst Ruska: electron microscopy

1986	Gerd Binnig, Heinrich Rohrer: scanning tunnelling microscopy
1987	Georg Bednorz, Alex Müller: ceramic superconductors
1991	Pierre-Gilles de Gennes: soft matter
1996	Robert Curl, Harold Kroto, Richard Smalley: fullerenes
1998	Walter Kohn: computational methods
2000	Alan Heeger, Alan MacDiarmid, Hideki Shirikawa: conductive polymers
2000	Zhores Alferov, Herbert Kroemer: semiconductor heterostructures
2000	Jack Kilby: integrated circuits
2007	Albert Fert, Peter Grünberg: giant magnetoresistance
2009	Charles Kao: optical fibres
2009	Willard Boyle, George Smith: charge-coupled devices
2010	André Geim, Konstantin Novoselov: graphene
2011	Dan Shechtman: quasicrystals

Materials

References

F. F. Abraham, 'How Fast Can Cracks Run? A Research Adventure in Materials Failure Using Millions of Atoms and Big Computers', *Advances in Physics* (2003), vol 52, pp 727–90.

Z. Alferov, 'Heterostructures for Optoelectronics: History and Modern Trends', *Proceedings of the IEEE* (2013), vol 101, pp 2176–82.

J.-F. Ayme, J. E. Beves, C. J. Campbell, and D. A. Leigh, 'Template Synthesis of Molecular Knots', *Chemical Society Reviews* (2013), vol 42, pp 1700–12.

J. Bardeen, 'Semiconductor Research Leading to the Point Contact Transistor', in *Nobel Lectures, Physics 1942–1962* (Amsterdam: Elsevier, 1964).

R. Barthes, *Mythologies* (New York: Noonday Press, 1972).

S. Beale, 'Precision Engineering for Future Propulsion and Power Systems: A Perspective from Rolls-Royce', *Philosophical Transactions of the Royal Society A* (2012), vol 370, pp 4130–53.

J. Bohr and K. Olsen, 'The Ancient Art of Laying Rope', *Europhysics Letters* (2011), vol 93, art 60004.

W. H. Bragg and W. L. Bragg, *X Rays and Crystal Structure* (London: Bell and Sons, 1915).

L. Brodsley, C. Frank, and J. W. Steeds, 'Prince Rupert's Drops', *Notes and Records of the Royal Society, London* (1986), vol 41, pp 1–26.

D. G. Cahill and R. O. Pohl, 'Lattice Vibrations and Heat Transport in Crystals and Glasses', *Annual Review of Physical Chemistry* (1988), vol 39, pp 93–121.

J. M. D. Coey, *Magnetism and Magnetic Materials* (Cambridge: Cambridge University Press, 2010).

J. Cross, 'The 100th Object: Solar Lighting Technology and Humanitarian Goods', *Journal of Material Culture* (2013), vol 18, pp 367–87.

J. M. Cullen and J. M. Allwood, 'Mapping the Global Flow of Aluminum: From Liquid Aluminum to End-Use Goods', *Environmental Science & Technology* (2013), vol 47, pp 3057–64.

P. P. Edwards, R. L. Johnston, C. N. R. Rao, D. P. Tunstall, and F. Hensel, 'The Metal-Insulator Transition: A Perspective', *Philosophical Transactions of the Royal Society A* (1998), vol 356, pp 5–22.

European Commission, *Critical Raw Materials for the EU* (Brussels: European Commission, 2010).

T. E. Graedel, E. M. Harper, N. T. Nassar, and B. K. Reck, 'On the Materials Basis of Modern Society', *Proceedings of the National Academy of Sciences* (2013), 2 December. doi: 10.1073/pnas.1312752110.

G. Grandin, *Fordlandia: The Rise and Fall of Henry Ford's Forgotten Jungle City* (New York: Metropolitan Books, 2009).

R. W. Gurney and N. F. Mott, 'The Theory of the Photolysis of Silver Bromide and the Photographic Latent Image', *Proceedings of the Royal Society A* (1938), vol 164, pp 151–67.

G. P. Hammond and C. Jones, 'Embodied Energy and Carbon in Construction Materials', *Proceedings of the Institution of Civil Engineers—Energy* (2008), vol 161, pp 87–98.

G. P. Hatch, 'Dynamics in the Global Market in Rare Earths', *Elements* (2013), vol 8, pp 341–6.

B. J. Hunt, 'The Ohm Is Where the Art Is: British Telegraph Engineers and the Development of Electrical Standards', *Osiris*, 2nd Series (1994), vol 9, pp 48–63.

H. N. Iben and J. F. O'Brien, 'Generating Surface Crack Patterns', *Graphical Models* (2009), vol 71, pp 198–208.

J. Jacobs, *The Economy of Cities* (New York: Random House, 1969).

C. K. Kao, 'Nobel Lecture: Sand from Centuries Past: Send Future Voices Fast', *Reviews of Modern Physics* (2010), vol 82, pp 2299–303.

L. Kristjánsson, *Iceland Spar and its Influence on the Development of Science and Technology in the Period 1780–1930*, 3rd edn (Institute of Earth Sciences, University of Iceland, 2010).

D. H. Meadows and others, *The Limits to Growth: A Report for the Club of Rome's Project on the Predicament of Mankind* (London: Universal Books, 1972).

J. L. Meikle, *American Plastic: A Cultural History* (New Brunswick: Rutgers University Press, 1995).

L. A. B. Pilkington, 'The Float Glass Process', *Proceedings of the Royal Society A* (1969), vol 314, pp 1–25.

A. Ramage and P. Craddock, *King Croesus' Gold: Excavations at Sardis and the History of Gold Refining* (London: British Museum Press, 2000).

W. Serrin, *Homestead: The Glory and Tragedy of an American Steel Town* (London: Vintage, 1993).

J. L. Simon, *The Ultimate Resource* (Princeton: Princeton University Press, 1981).

C. S. Smith, 'The Discovery of Carbon in Steel', *Technology and Culture* (1964), vol 5, pp 149–75.

W. Thomson, 'On the Electric Conductivity of Commercial Copper of Various Kinds', *Proceedings of the Royal Society of London* (1856–1857), vol 8, pp 550–5.

L. R. G. Treloar, *The Physics of Rubber Elasticity*, 3rd edn (Oxford: Clarendon Press, 2005).

G. Tweedale, 'Sir Robert Hadfield F.R.S. (1858–1940), and the Discovery of Manganese Steel', *Notes and Records of the Royal Society of London* (1985), vol 40, pp 63–74.

Further reading

Chapter 1: Gold, sand, and string

A. Forty, *Concrete and Culture: A Material History* (London: Reaktion, 2012).

J. Goody, *Metals, Culture and Capitalism* (Cambridge: Cambridge University Press, 2012).

T. J. Misa, *A Nation of Steel: The Making of Modern America, 1865–1925* (Baltimore: Johns Hopkins University Press, 1995).

G. Riello, *Cotton: The Fabric that Made the Modern World* (Cambridge: Cambridge University Press, 2013).

G. Tweedale, *Steel City: Entrepreneurship, Strategy and Technology in Sheffield, 1743–1993* (Oxford: Clarendon Press 1995).

R. F. Tylecote, *A History of Metallurgy*, 2nd edn (London: Institute of Metals, Maney, 1992).

O. H. Wyatt and D. Dew-Hughes, *Metals, Ceramics and Polymers* (Cambridge: Cambridge University Press, 1974).

Chapter 2: Close inspection

H. K. D. H. Bhadeshia and R. Honeycombe, *Steels: Microstructure and Properties*, 3rd edn (London: Butterworth-Heinemann, 2006).

M. Born and K. Huang, *Dynamical Theory of Crystal Lattices* (Oxford: Clarendon Press, 1956).

R. W. Cahn, *The Coming of Materials Science* (Amsterdam: Pergamon, 2001).

P. P. Ewald (ed), *Fifty Years of X-Ray Diffraction* (Utrecht: N.V.A. Oosthoek for the International Union of Crystallography, 1962).

L. Hoddeson, E. Braun, J. Teichmann, and S. R. Weart (eds), *Out of the Crystal Maze: Chapters from the History of Solid-State Physics* (Oxford, 1992).

Chapter 3: Tough but slippery

J. Akhavan, *The Chemistry of Explosives*, 3rd edn (Cambridge: Royal Society of Chemistry, 2011).

M. F. Ashby, *Materials Selection in Mechanical Design*, 3rd edn (Oxford: Butterworth-Heinemann, 2005).

M. Doi, *Soft Matter Physics* (Oxford: Oxford University Press, 2013).

P.-G. de Gennes and J. Badoz, *Fragile Objects* (New York: Copernicus, 1996).

G. Grimvall, *Thermophysical Properties of Materials*, 2nd edn (Amsterdam: North Holland, 1999).

R. Hill, *The Mathematical Theory of Plasticity* (Oxford: Clarendon Press, 1950).

K.-E. Kurrer, *The History of the Theory of Structures: From Arch Analysis to Computational Mechanics* (Berlin: Ernst & Sohn, 2008).

D. Tabor, *The Hardness of Metals* (Oxford: Clarendon Press, 2000).

S. P. Timoshenko, *History of Strength of Materials* (New York: Dover, 1983).

Chapter 4: Electric blue

M. Fox, *The Optical Properties of Solids*, 2nd edn (Oxford: Oxford University Press, 2010).

R. A. LeSar, *Introduction to Computational Materials Science* (Cambridge: Cambridge University Press, 2013).

J. W. Orton, *The Story of Semiconductors* (Oxford: Oxford University Press, 2008).

L. Solymar, D. Walsh, and R. R. A. Sims, *The Electrical Properties of Materials*, 9th edn (Oxford: Oxford University Press, 2014).

R. J. D. Tilley, *Colour and Optical Properties of Materials*, 2nd edn (Chichester: Wiley, 2011).

A. S. Travis, *The Rainbow Makers: The Origins of the Synthetic Dyestuffs Industry in Western Europe* (Bethlehem, PA: Lehigh University Press, 1993).

Materials

Chapter 5: Making stuff and making things

E. Braun and S. Macdonald, *Revolution in Miniature: The History and Impact of Semiconductor Electronics*, 2nd edn (Cambridge: Cambridge University Press, 1982).

J. B. Pendry, 'Negative Refraction', *Contemporary Physics* (2004), vol 45, pp 191–202.

R. C. Reed, *The Superalloys: Fundamentals and Applications* (Cambridge: Cambridge University Press, 2006).

Chapter 6: Such quantities of sand

J. M. Allwood, J. M. Cullen, M. A. Carruth, D. R. Cooper, M. McBrien, R. L. Milford, M. Moynihan, and A. C. H. Patel, *Sustainable Materials: With Both Eyes Open* (Cambridge: UIT, 2012).

K. Gill, *Of Poverty and Plastic: Scavenging and Scrap Trading Entrepreneurs in India's Urban Informal Economy* (Oxford: Oxford University Press, 2010).

G. Gunn, *Critical Metals Handbook* (Chichester: Wiley, 2014).

D. Miller, *Stuff* (Cambridge: Polity Press, 2010).

D. A. Singer, and W. D. Menzie, *Quantitative Mineral Resource Assessments* (Oxford: Oxford University Press, 2010).

J. Tully, *The Devil's Milk: a Social History of Rubber* (New York: Monthly Review Press, 2011).

United Nations Environment Programme, *Metal Stocks in Society* (UNEP, 2010).

Index

Expand your collection of
VERY SHORT INTRODUCTIONS